I0464525

[H.A.S.C. No. 113–119]

UNMANNED CARRIER–LAUNCHED AIRBORNE SURVEILLANCE AND STRIKE (UCLASS) REQUIREMENTS ASSESSMENT

————

HEARING

BEFORE THE

SUBCOMMITTEE ON SEAPOWER AND PROJECTION FORCES

OF THE

COMMITTEE ON ARMED SERVICES
HOUSE OF REPRESENTATIVES

ONE HUNDRED THIRTEENTH CONGRESS

SECOND SESSION

————

HEARING HELD
JULY 16, 2014

————

U.S. GOVERNMENT PUBLISHING OFFICE

89–512   WASHINGTON : 2015

For sale by the Superintendent of Documents, U.S. Government Publishing Office
Internet: bookstore.gpo.gov   Phone: toll free (866) 512–1800; DC area (202) 512–1800
Fax: (202) 512–2104   Mail: Stop IDCC, Washington, DC 20402–0001

CONTENTS

CHRONOLOGICAL LIST OF HEARINGS

2014

WEDNESDAY, JULY 16, 2014

UNMANNED CARRIER–LAUNCHED AIRBORNE SURVEILLANCE AND STRIKE (UCLASS) REQUIREMENTS ASSESSMENT

STATEMENTS PRESENTED BY MEMBERS OF CONGRESS

WITNESSES

APPENDIX

Page

UNMANNED CARRIER–LAUNCHED AIRBORNE SURVEILLANCE AND STRIKE (UCLASS) REQUIREMENTS ASSESSMENT

HOUSE OF REPRESENTATIVES,
COMMITTEE ON ARMED SERVICES,
SUBCOMMITTEE ON SEAPOWER AND PROJECTION FORCES,
Washington, DC, Wednesday, July 16, 2014.

The subcommittee met, pursuant to call, at 2:00 p.m., in room 2212, Rayburn House Office Building, Hon. J. Randy Forbes (chairman of the subcommittee) presiding.

OPENING STATEMENT OF HON. J. RANDY FORBES, A REPRESENTATIVE FROM VIRGINIA, CHAIRMAN, SUBCOMMITTEE ON SEAPOWER AND PROJECTION FORCES

Mr. FORBES. We want to welcome you to this hearing this afternoon. I apologize at the beginning; we are going to have a vote series that takes place, but we will be back. It is an important hearing and we want to go as long as it takes to get this done.

Today the subcommittee convenes to receive testimony on the Unmanned Carrier-Launched Airborne Surveillance and Strike (UCLASS) program.

Our first panel of distinguished guests testifying before us are Mr. Ronald O'Rourke. He is a specialist in naval affairs, Defense Policy and Arms Control Section for the Congressional Research Service; Mr. Robert Martinage, former Deputy Under Secretary of the Navy; Mr. Shawn Brimley, Executive Vice President and Director of Studies for the Center for a New American Security; and Mr. Bryan McGrath, Managing Director of FerryBridge Group, LLC.

Gentlemen, thank you so much for being here.

Collectively, this bipartisan group has advised the United States Congress and Presidential campaigns, commanded a Navy large-surface combatant, drafted the 2007 maritime strategy, served as Under Secretary of the Navy, served on the National Security Council staff, and worked at various distinguished think tanks.

Given their diverse background, I am confident that this bipartisan group of witnesses will be able to provide a detailed perspective of this committee's continued work on the UCLASS program.

Our second distinguished panel, which will immediately follow this one, includes Navy and Joint Staff leaders, including Vice Admiral Paul A. Grosklags, Principal Military Deputy, Assistant Secretary of the Navy for Research, Development and Acquisitions; Mr. Mark Andress, Assistant Deputy Chief of Operations for Information Dominance; Brigadier General Joseph Guastella, Director of Joint Requirements Oversight Council, Department of Defense.†

Gentlemen, once again, we thank you for being here.

We have called this hearing to discuss the Navy's UCLASS program. But before we proceed, I want to be clear from the onset that I am a strong supporter of a future carrier air wing that is comprised of both manned and unmanned aviation assets. The F/A–18 Super Hornet, the F–35C, the EA–18G Growler, the E–2 Hawkeye, and the UCLASS program will all be integral to ensuring our carrier fleet can continue to project power throughout the globe.

I believe the fundamental question we face is not about the utility of unmanned aviation to the future air wing, but the type of unmanned platform that the UCLASS program will deliver and specific capabilities this vital asset will provide the combatant commander.

Given the likely operational environment of the 2020s and beyond, including in both the Western Pacific Ocean and Persian Gulf, I believe strongly that the Nation needs to procure a UCAV [unmanned combat air vehicle] platform that can operate as a long-range surveillance and strike asset in the contested and denied A2/AD [anti-access/area-denial] environments of the future.

Unfortunately, in its current form, this committee has concluded the UCLASS air systems segments requirements will not address the emerging anti-access/area-denial challenges to U.S. power projection that originally motivated creation of the Navy Unmanned Combatant Air System program during the 2006 Quadrennial Defense Review [QDR] and which were reaffirmed in both the 2010 QDR and 2012 Defense Strategic Guidance.

It is my determination that the disproportionate emphasis in the requirements on unfueled endurance to enable continuous intelligence, surveillance, and reconnaissance [ISR] support to the carrier strike group would result in an aircraft design that would have serious deficiencies in both survivability and internal weapons payload capacity and flexibility.

Furthermore, the cost limits for the aircraft are more consistent with a much less capable aircraft and will not enable the Navy to build a relevant vehicle that leverages readily available and mature technology.

In short, developing a new carrier-based manned aircraft that is primarily another unmanned ISR sensor that can operate in a medium- to high-level threat environment would be a missed opportunity and inconsistent with the 2012 Defense Strategic Guidance, which called for the United States to maintain its ability to project power in areas in which our access and freedom to operate are challenged.

But the question of UCLASS is not just one of design and capability. It is also about the roll and responsibility that Congress has in cultivating, supporting, and protecting military innovation.

Like with the shift from cavalry to mechanized forces, sailing ships to steam-powered vessels, the battleship to naval aviation, or adopting unmanned aerial vehicles in the late 1990s, ideas that ini-

---

† Mr. Forbes corrected his remarks for the record and recognizes General Guastella's correct title, "Deputy Director for Requirements (J-8), Joint Staff."

tiate difficult changes and disrupt current practices are often first opposed by organizations and bureaucracies that are inclined to preserve the status quo.

I believe the Congress has a unique role to help push the Department and the services in directions that, while challenging, will ultimately benefit our national security and defense policy.

I therefore intend to use this hearing today to explore not just the UCLASS program, but the broader utility a UCAV can have on the Navy's ability to continue to project power from the aircraft carrier and the implications for the power-projection mission in the future if we proceed down the current course.

Again, I thank our two panels for being here to testify and look forward to your testimony.

[The prepared statement of Mr. Forbes can be found in the Appendix on page 45.]

Mr. FORBES. And with that, I turn to my good friend, Mr. Courtney, for any comments he might have.

STATEMENT OF HON. JOE COURTNEY, A REPRESENTATIVE FROM CONNECTICUT, COMMITTEE ON ARMED SERVICES

Mr. COURTNEY. Great. Thank you, Mr. Chairman.

I want to thank the panel for being here. Given the time squeeze, I am going to be very brief.

Reading the testimony of both panels, I actually think that—you know, really, I think everyone is trying to get to the same end result here, which is a carrier-based unmanned air wing. I think there is important discussion that needs to take place about sort of the path in terms of moving forward.

And, again, I think, even though we are sort of walking a tightrope here a little bit because we are talking about a classified process, so we really can't fully flesh out, I think, all aspects of that path at this hearing because it is a public hearing, not a classified hearing. Again, I look forward to the testimony.

Again, Mr. McIntyre had some brief remarks which, again, for the record, I would ask that they be entered.

Mr. FORBES. Without objection, we will put any comments that Mr. McIntyre has in the record.

[The prepared statement of Mr. McIntyre can be found in the Appendix on page 50.]

Mr. COURTNEY. With that, I will yield back.

Mr. FORBES. Thank you, Joe.

And, with that, Mr. O'Rourke, we would love to hear any comments that you might have.

STATEMENT OF RONALD O'ROURKE, SPECIALIST IN NAVAL AFFAIRS, CONGRESSIONAL RESEARCH SERVICE

Mr. O'ROURKE. Chairman Forbes, Ranking Member Courtney, and distinguished members of the subcommittee, thank you for the opportunity to appear before you today to testify on the UCLASS program.

Mr. Chairman, with your permission, I would like to submit my written statement for the record and summarize it here briefly.

Mr. FORBES. And, without objection, all the statements of our witnesses will be submitted for the record.

Mr. O'ROURKE. As requested, my testimony identifies some issues the subcommittee might consider in assessing operational requirements for the UCLASS program.

My statement presents six such issues. The first is whether we are currently undergoing a shift in strategic eras.

World events since late last year have led to a discussion among observers about whether we are currently shifting from the familiar post-Cold War era of the last 20 to 25 years to a new and different strategic area characterized by, among other things, renewed great power competition.

The shift from the Cold War to the post-Cold War era led to a reassessment of assumptions and frameworks of analysis regarding defense funding levels, strategy and missions that resulted in numerous changes in U.S. defense plans and programs while leaving other programs unchanged. A shift from the post-Cold War era to a new strategic era could lead to another such reassessment.

Current requirements for the UCLASS program reflect analyses that were done between 2009 and 2011 and then updated and revalidated from 2012 through April 2013. This activity predates the events starting in late 2013 that have led to the discussion over the possible shift in strategic eras.

Potential questions include the following:

First, are we undergoing a shift from the post-Cold War era to a new strategic era?

Second, if we are undergoing such a shift, should that lead to a reassessment of assumptions and frameworks of analyses relating to defense funding levels, strategy, and missions?

And, third, if there is such a reassessment, what effect, if any, might it have on UCLASS requirements?

A second issue the subcommittee might consider is how requirements for the UCLASS program might affect cost, schedule, and technical risk.

On the issue of cost, the Navy explained to me that the program's affordability KPP [key performance parameters] is based on the UCLASS AOA [analysis of alternatives] update and Navy discussions with industry about potential costs for the UCLASS program as currently defined, plus lessons from the UCAS–D [Unmanned Combat Air System Demonstrator program] effort.

Defining the affordability KPP in this manner can help ensure that the affordability KPP is realistic for the program as currently defined.

At the same time, in the context of a debate over requirements, this approach can produce a definition of affordability that can be viewed as circular, to some degree, because it can be understood as saying, in essence, what is affordable is the program with the current requirements.

A definition of affordability that is, to some degree, circular in nature in relation to requirements has the potential for being invoked as a rhetorical device for discouraging or closing down debate on requirements.

A third issue the subcommittee may wish to consider is how requirements for the UCLASS program might affect estimated outcomes in future operational scenarios.

The specific tactical situations that were examined in the UCLASS AOA are related to the program's current requirements. Assessing alternative requirements could involve examining potential outcomes in other tactical situations, and a broader analysis might examine how changes in requirements might affect estimated outcomes in campaign-level force-on-force situations rather than in specific tactical situations.

A fourth issue the subcommittee might consider is how UCLASS requirements relate to assessments of potential future adversary capabilities, for example, how sensitive are requirements for the UCLASS program to changes and assessments of potential future adversary capabilities and how much uncertainty or potential for changes is there in these threat assessments.

A fifth issue the subcommittee might consider is how requirements for the UCLASS program might affect potential technology paths for future systems and capabilities, for example, what effect might UCLASS requirements have on opening up, preserving, or encumbering potential pathways for achieving the Navy's current long-term vision for naval aviation or potential alternatives to that vision.

A sixth issue the subcommittee might consider is how requirements for the UCLASS program might affect the behavior of other countries. For example, what impact might UCLASS requirements have in terms of imposing costs on potential adversaries or persuading potential adversaries—dissuading potential adversaries from taking certain courses of action or reassuring U.S. allies and partners regarding U.S. intentions and resolve.

These six issues are by no means the only ones that might be raised, but considering them might help in forming a framework of analysis for assessing UCLASS requirements.

Mr. Chairman, that concludes my remarks, and I look forward to the subcommittee's questions.

[The prepared statement of Mr. O'Rourke can be found in the Appendix on page 51.]

Mr. FORBES. Thank you, Mr. O'Rourke.

Mr. Martinage, we look forward to your comments. Thank you for being here.

STATEMENT OF ROBERT MARTINAGE, SENIOR FELLOW, CENTER FOR STRATEGIC AND BUDGETARY ASSESSMENTS

Mr. MARTINAGE. Chairman Forbes, Ranking Member McIntyre, and members of this distinguished committee, first off, thank you for the opportunity to share my views on system performance requirements for UCLASS.

I would also like to express my appreciation to the committee for taking an active interest in what is one of the most important force development issues facing the Department of Defense and the Navy in particular.

I really don't think it is much of an exaggeration to say that what is at stake here is not just the operational relevance of the carrier air wing in the future, but, really, the strategic relevance of the aircraft carrier for decades to come.

I would like to highlight four themes from my written statement: first, how to think about UCLASS requirements broadly; second, the opportunity cost of unnecessarily high unrefueled endurance; third, some thoughts about payload requirements; and, fourth, what a more balanced UCLASS design might look like.

So, first and foremost, an assessment of UCLASS requirements should begin with a very simple question: What is the core operational challenge that UCLASS should be designed to solve?

The dominant answer within the Navy currently and reportedly reflected in the UCLASS draft request for proposal, or RFP, is that UCLASS is needed to maintain continuous maritime domain awareness around the carrier strike group as well as to identify targets for attack by relatively short-range manned fighters.

An alternative view, and one that reaches back to the initiation of the program by OSD [Office of the Secretary of Defense] and then the CNO [Chief of Naval Operations] Roughhead in 2009, is that the more pressing problem is maintaining our ability to project power from the sea when, one, carriers are compelled to stand off a considerable distance, perhaps 1,000 miles or more, from an adversary's territory due to emerging anti-access and area-denial challenges, like anti-ship ballistic missiles, anti-ship cruise missiles, wake-homing torpedos—and the list goes on and on—and then, second, when it is necessary to find and attack fixed and relocatable targets that are defended by modern innovative air defense systems.

If you believe we need more capacity to generate maritime domain awareness around the carrier strike group than what will be available when more than 60 MQ–4C Tritons, formerly BAMS [Broad Area Maritime Surveillance], enter into service, along with MQ–8B/C Fire Scouts that can operate off any air-capable ship in the fleet, then the current draft RFP, at least as reported in the press, is probably about right.

If you believe we need even more capacity for persistent ISR and light strike in low-to-medium threat environments, beyond the several hundred aircraft and the Predator, Gray Eagle, and Reaper fleets, then the draft RFP for UCLASS is probably on track.

If you believe instead that UCLASS should be the next step in the evolution of the carrier air wing and must be able to provide sea-based surveillance and strike capacity in anticipated anti-access and area-denial environments, then the Navy is aiming well off the mark, which brings us to theme two: The opportunity cost of the current threshold requirement for unrefueled endurance.

Driven by the perceived need to maintain continuous maritime domain awareness around the carrier strike group, including overnight while the deck is closed, the draft RFP reportedly contains a derived threshold requirement for an unrefueled endurance of about 14 hours.

The opportunity cost of that 14 hours of unrefueled endurance, however, are permanent aircraft design trades that reduce survivability and payload carriage and flexibility, the exact same attributes that are needed to perform ISR and precision strike in an anti-access/area-denial environment.

I would like to stress that these reductions in survivability and payload cannot be bought back later or added to future UCLASS variants.

Similarly, claims that threshold growth or objective requirements will place competitive pressure on industry to enhance survivability and payload attributes are mostly smoke and mirrors. They may appear compelling, but they are misleading.

As a matter of physics, absence breakthrough in engine technology, it is impossible to achieve 14 hours of unrefueled endurance with an air vehicle sized to operate from the aircraft carrier without making changes to its shape and propulsion path that negatively impact radar cross-section reduction, a.k.a [also known as] stealth, and reduce internal weapons carriage capacity, meaning both numbers and types of weapons that the air vehicle can carry.

Simply put, meeting the threshold requirement of 14 hours of unrefueled endurance necessarily results in sacrificing survivability, weapons carriage/flexibility and the number of weapons you can carry, and growth margins for future mission payloads. And, again, there are no technologically viable growth paths for restoring these attributes later.

Perhaps this opportunity cost would be acceptable if there was a compelling operational justification for 14 hours of unrefueled endurance, but there is not.

And the aircraft with 8 to 10 hours of unrefueled endurance flying at high subsonic speeds would have roughly three times the combat radius of F–18E/F or the F–35C.

So to put that into operational perspective, that same 8- to 10-hour endurance aircraft could launch from a carrier positioned 1,000 miles away from an area of interest, which happens to be the range of the Chinese DF–21D anti-ship ballistic missile, loiter on station for 3 to 4 hours, then recover onboard the carrier still with gas in the tank.

When factoring in aerial refueling, which is typically available in wartime, the 14-hour unrefueled endurance requirement is even more nonsensical. With refueling, that same 8- to 10-hour endurance aircraft could remain aloft for 24 to 48 hours or longer.

I would like to shift now to the third theme, payload requirements. I am not aware of any mission or campaign-level analysis that supports a payload requirement of 1,000 pounds for a carrier-based strike aircraft. Certainly that is not the case with either the F–18 or the F–35.

Put more plainly, 1,000 pounds of payload, which equates to four small-diameter bombs, is clearly inadequate for saturating an adversary's short-range air defenses and neutralizing a wide range of very relevant target sites, such as coastal defense cruise missile sites, air defense radars, missile launchers, even enemy service combatants. One thousand pounds of strike payload per aircraft just isn't enough.

In addition, scant consideration appears to have been given to the types of weapons that UCLASS should be able to accommodate.

Even a stealthy UCLASS in the future will need to stand off from some classes of defended targets. So it should be able to carry weapons such as the Joint Standoff Weapon, the Long Range Anti-Ship Missile, or LRASM, and/or a Joint Strike Missile. All of these things need more consideration.

So now, for my fourth and final theme: What would a more balanced UCLASS design look like? A more balanced carrier-based unmanned air vehicle would, first, achieve the minimum level of signature reduction required to locate priority targets and engage them with available weapons without being destroyed by modern air defenses or, put another way, needs to be able to find and hit targets without being shot down. That is a minimum precondition.

Second, it needs sufficient unrefueled endurance to reach target areas when carriers are forced to stand off 1,000 miles or more.

And then, third, once those two conditions are met—it can find and hit targets without being shot down, it has a meaningful operational combat radius—the next thing is maximizing the amount of payload it can carry and as many types of weapons that it can carry while still fitting on the carrier deck.

So using that approach, a carrier-based UAS [unmanned aircraft system] in the future could have, for example, an unrefueled endurance of 8 to 10 hours, which translates to a combat radius of 1,700 to 2,000 nautical miles, either from a carrier or from a tanker; 24 to 48 hours of mission endurance with air-to-air refueling; broadband/all-aspect, radar cross-section reduction matched to the anticipated threat environment of 2025 and beyond; and the ability to carry 3- to 4,000 pounds of strike payload internally, roughly what an F–35C can carry, including their variety of direct and standoff weapons.

With those attributes, a balanced UCLASS could serve as an independent, long-range surveillance and striking arm of the aircraft carrier in anti-access and area-denial environments.

With aerial tanking support, it could respond globally to short-notice aggression, regardless of the carrier's initial location, and contribute to a sustained extended-range precision strike campaign against an adversary's fixed and mobile target as part of a joint force.

So to conclude and to just foot-stomp a few points, first, the opportunity cost of 4 to 6 hours of additional unrefueled endurance, so 14 hours vice 8 to 10, is a dramatic reduction in strike capacity and flexibility, a significant increase in air vehicle vulnerability and reduced growth potential, meaning lower margins for space, weight, power, and cooling.

Second, be very skeptical of growth paths that promise to increase survivability and payload later. Yes. There are band-aid solutions and some workarounds, but the core design trades made to achieve 14 hours of unrefueled endurance involve the air vehicle's shape and propulsion path, and they cannot be reversed, period.

There is no question that the Nation needs a carrier-based unmanned aircraft. The relevant question is what kind of aircraft. The air vehicle called for in the UCLASS RFP appears to be optimized for sustaining persistent maritime domain awareness——

Mr. FORBES. I am going to have to interrupt you there because we have got votes that are called. We will let you wrap up very briefly when we get back and then go right to Mr. Brimley and Mr. McGrath.

We apologize for these votes. Unfortunately, we are looking at probably about 3:45 before we will be back. So we are going to stand in recess until that time.

[Recess.]

Mr. FORBES. Thank you once again for your patience. And we apologize for these votes.

Mr. Martinage, I think you were finishing up. If you could take about 60 seconds and wrap up, and then we will move on to Mr. Brimley.

Mr. MARTINAGE. As I was saying, there is no question that the Nation needs a carrier-based unmanned aircraft. The relevant question is what kind of aircraft.

And, in my view, the air vehicle called for in the UCLASS RFP appears to be optimized for sustaining persistent maritime domain awareness and ISR coverage in relatively benign threat environments.

And, in my view, that is redundant with aircraft in service or soon to be in service in the Navy, the Army, and the Air Force.

And, most critically, it does not address the core operational problem facing aviation: The intensifying anti-Navy threats that will push the carrier farther away from target areas and network air defenses that will make non-stealthy aircraft increasingly vulnerable to detection and attack. And that is the problem we need to look at.

And I look forward to your questions and discussions later on.

[The prepared statement of Mr. Martinage can be found in the Appendix on page 61.]

Mr. FORBES. Well, thank you for your comments.

Mr. Brimley, look forward to your comments.

STATEMENT OF SHAWN BRIMLEY, EXECUTIVE VICE PRESIDENT AND DIRECTOR OF STUDIES, CENTER FOR A NEW AMERICAN SECURITY

Mr. BRIMLEY. Thank you, Chairman Forbes, Ranking Member Courtney, for the opportunity to testify.

I want to acknowledge my co-panelists, whose work I very much admire.

I see the issue of how the Navy approaches the Unmanned Carrier-Launched Airborne Surveillance and Strike program, or UCLASS, as an important indicator of how serious the Department of Defense is in ensuring America's long-term military technical advantage.

I am concerned that the current program does not fully exploit the opportunity the Navy has, in my mind, to lock in what could be a decisive advantage in future warfare, the ability to employ long-range, stealthy, unmanned strike platforms from the aircraft carrier.

As a former civilian who worked national security policy at both OSD, the Office of Secretary of Defense, at the White House, I typically approach design procurement—defense procurement and design issues through the lens of a policymaker and I ask the following types of questions:

Number one, will the platform provide a future Commander in Chief better military options during a crisis?

Two, will it help address pressing gaps in U.S. defense strategy and planning?

Three, does it enable forward U.S. forces to present a stronger conventional deterrent and, if necessary, help ensure U.S. forces can defeat a plausible adversary?

Number four, will the program help underwrite the confidence of our allies and partners?

Five, does it reflect measured judgments regarding mid- to long-term requirements for U.S. defense?

And, six, does the program help ensure America's military technical dominance in an increasingly competitive environment?

Having followed as best I can the debate surrounding UCLASS program, I am concerned that the answers to most of the questions I just outlined are "no."

The specific requirements in the current draft request for proposals, in my mind, having read as much of the open-source material as I can, will result in a platform that, one, fails to add any real striking power to the carrier air wing; two, duplicates many of the ISR systems already available to the Navy; three, does nothing to address the major threat facing the aircraft carrier, the need to operate from longer ranges due to improvements in anti-ship, ballistic and cruise missile design; four, and most problematically, vectors the Navy down an investment path that will waste precious time and money, in my view, risking our ability to integrate long-endurance, strike-capable unmanned systems into this country's most important power-projection asset, the aircraft carrier.

I think the strategic implications of a failure to push hard now to develop carrier-launched unmanned combat aerial vehicles could be significant. Budgets are tight and hard choices must be made, but this is an area where I don't think we can afford to get it wrong.

To do so will end up costing more money over the long term and increase the risk that the U.S. Navy and the broader joint force will be ill-prepared for important plausible future contingencies.

In this respect, I fully endorse, Mr. Chairman, what this committee did in requiring the Secretary of Defense—and I think it is important it be the civilian leadership of the Department—certify the requirements for this program before further substantial funding is committed.

This committee adjudicates issues involving programs much larger and far more costly than the UCLASS program, but I think this is one of those rare decisions regarding setting requirements for future capabilities that could have a major impact on how tomorrow's joint force might fight a future war.

It is critical, in my view, to take the time to ensure that we get this right. I appreciate being invited to speak today and look forward to the discussion.

[The prepared statement of Mr. Brimley can be found in the Appendix on page 73.]

Mr. FORBES. Thank you, Mr. Brimley, for your comments.

Mr. McGrath.

STATEMENT OF BRYAN MCGRATH, MANAGING DIRECTOR, FERRYBRIDGE GROUP, LLC

Mr. McGRATH. Thank you, Chairman Forbes, Ranking Member Courtney. Thank you for the opportunity to be with you today.

Thank you for your leadership in sustaining the competitive advantages of American seapower and for the leadership that you have exerted thus far in exerting pressure on the Department of Defense to ensure a truly capable Unmanned Carrier-Launched Surveillance and Strike system.

It is an honor to be on the panel with the three gentlemen here, who are also good friends. Bob Martinage, Shawn Brimley, and Ronald O'Rourke are among the smartest thinkers on the scene today and to be counted among them is humbling.

And while Mr. O'Rourke's background and employment preclude political or ideological identification, I think it is noteworthy to note the presence of two Obama administration political appointees, Mr. Martinage and Mr. Brimley, alongside me, and myself, the Navy policy team co-lead for the 2012 Romney for President Committee.

The fact that the three of us are in solid agreement on the need for continued congressional oversight of the Navy's UCLASS acquisition is notable.

Specifically, there appears to be consensus on the need to ensure the Navy does not pursue a largely duplicative system that does little to advance the striking power of our Nation's primary forward-deployed power-projection system, the aircraft carrier strike group.

I believe we have reached a ''for want of a nail, a kingdom is lost'' moment. The aircraft carrier has been—its demise has been predicted for 60 years. And that demise hasn't happened because its air wing has evolved to pace the threat throughout its history. It is agnostic to the weapons it projects.

If the air wing of the future does not evolve in a way that enables the kind of unmanned strike that a truly capable UCLASS would bring, the aircraft carrier might indeed become obsolescent.

If it becomes obsolete, the preponderant Navy that we field today that is the primary—in my view, the primary sustainer of the global system that is in place today will become far less powerful. Far less powerful and influential Navy means a far less powerful and influential United States.

This is not a small question. It is a large one. And I appreciate your leadership on the subject. Thank you.

[The prepared statement of Mr. McGrath can be found in the Appendix on page 86.]

Mr. FORBES. Thank you, Mr. McGrath.

And now we would like to—I am going to defer my questions until after Mr. Courtney.

So, Mr. Courtney, I will let you go first if you have any questions.

Mr. COURTNEY. Thank you, Mr. Chairman.

And, again, I think this is an important hearing. Obviously, it was something that was part of the House Defense Authorization bill.

I feel a little bit like we are shadowboxing here, though, because we are talking about a classified RFP process.

And, again, I think the latter two witnesses—your remarks were, I think, at a level that I think comport with that because you are talking about, you know, the long-range mission—or goal of this program, which is—again, I think it is great to have that discussion.

You know, Mr. Martinage, I mean, some of your comments were really focused on very specific, you know, components. I mean, the 1,000-gallon fuel item that you mentioned a couple of times in your remarks.

And, I mean, again, just for the record, I mean, from a process standpoint, we are in a place right now where there is, again, a classified RFP that is going to be going out in the next few months or so.

Have you seen any of those documents that, you know, provide the basis for your testimony today?

Mr. MARTINAGE. I have not seen the final draft RFP or the most recent RFP. I have been paying very close attention to the materials that are out about the draft RFP in the public domain as well as the KPPs and KSAs [key system attributes] discussion, which has been pretty extensive in the public domain.

And I think the issue—I don't know—which is the central one, I think, in my testimony, about the 14 hours of unrefueled endurance is clearly a parameter that is out there in the public domain.

And the opportunity cost of that 14 hours of unrefueled endurance clearly has an opportunity cost in both survivability and payload, and that is just a matter of physics.

It is not a classification issue. It is not a sensitivity issue. It is an aircraft design issue. And given where we are with engine propulsion technology, you just can't get to 14 hours unless you do things to the shape of the aircraft and the propulsion path that compromise stealth and payload.

And that is why I think the current path we are on is not a balanced design. And if you relaxed that threshold requirement for unrefueled endurance, you could dramatically improve payload and survivability, and that is my point.

Mr. COURTNEY. So, Mr. O'Rourke, I mean, in the past, I mean, we have had weapons platforms—excuse me—and systems that have started out looking one way and then, over time, have evolved or adapted to different capabilities and different—maybe you could just give some historic perspective in terms of other programs that adaptation and evolution has occurred.

Mr. O'ROURKE. Yeah. To just pick a few examples that come to mind in the area of carrier-based aviation, the F–18 is probably the largest single example.

It went through multiple versions, from the AB to the CD and then to the larger version, the EF, the Super Hornet, and off of that they also then developed the EA–18G.

Another example would be the E–2 Hawkeye and how it has evolved from the E–2C to the Hawkeye 2000 to the E–2D Advanced Hawkeye with its new radar.

A third example would be the P–3 that has evolved a number of times since the 1960s through a series of updates. And the Navy's plan right now is to procure the P–8 Poseidon multi-mission maritime aircraft with an incremental or step upgrade in mind.

So right now that plane exists at something called Increment 1, but there are plans to build it in a new version called Increment 2 that will implement three different engineering change proposals over the next few years and then move on beyond that to something called Increment 3. And all these things are supposed to [achieve] IOC [initial operating capability] between fiscal year 2014 and fiscal year 2020.

In general, it is worth noting that spiral development, which I think is the idea that you are getting at here, is established as an acquisition pathway for DOD programs. And, in fact, there was a push several years ago to make it the default approach to acquisition for DOD programs.

Mr. COURTNEY. Right.

And, I mean, even other sort of non-aviation—I mean, DDGs [guided missile destroyers] have also kind of changed their look over the years.

Mr. O'ROURKE. That is right.

In the area of shipbuilding, there are additional examples. You mentioned one, the DDG, which has moved from the Flight I to the Flight II, from there to the Flight IIA, and now we are planning on shifting to the Flight III.

The 688-class submarine went through a number of changes, and the 688s we built at the end of that program are quite different from the early ones. And the *Virginia* class is going through a block upgrade.

So, yes, this same idea is well established in shipbuilding as well.

Mr. COURTNEY. So, I mean—so I guess—and I don't have much more to ask right now.

Is that—I mean, that is sort of the question of the hearing, you know, really, whether or not, you know, this is a fork in the road that is irrevocable and, you know, permanent, forever, or whether or not, you know, that we can follow other precedents in the past.

And certainly, when the next panel comes up, that certainly would be my question that I would certainly want to pose to them. And, you know, I may have some other written questions afterwards.

But, you know, with that, Mr. Chairman, I would just yield back to you.

Mr. FORBES. Thank you, Mr. Courtney.

Also, we have a number of our Members who would like to submit written questions who couldn't be here because of the votes.

I would like to walk through a series of questions, if I could.

And, Mr. O'Rourke, I would like to start with you because you are kind of the closest we have to a historian in looking at this from the Congressional Research Service.

Do you see any shift in the international security environment that might warrant the Navy pursuing a different path on the UCLASS program than what it is currently pursuing?

Mr. O'ROURKE. That is the first of the six issues I raise, whether we are undergoing right now a shift in strategic eras. There are a number of people who feel that we are undergoing such a shift.

My own personal view, as an analyst, is that, yes, I think we are experiencing a shift in strategic eras. Right now I am watching that situation and have been for a number of months.

Mr. FORBES. Could you give me a couple of examples of the world situation that would justify you making that comment?

Mr. O'ROURKE. Well, I think what caught the attention of the people who have written about this potential shift in strategic eras are two sets of developments.

One are those in the Western Pacific, and that has to do with a series of actions by China starting late last year that appear intended or aimed at gaining a greater degree of Chinese control over its near-seas region.

That included the announcement of the air defense identification zone toward the end of November, the incident with the *Cowpens* in December, the imposition of the fishing regulations in January, and then most recently the movement of the oil rig to the Paracel Islands starting in May.

So that is one-half of the situation that I think a number of these observers were noticing.

And the other was Russia's seizure and annexation of Crimea, which was a landmark event regarding the division of territories within Europe since the end of World War II.

And the observers who have looked at that have said, in essence, that we may be shifting to a new strategic era, that the unipolar moment, as it were, is over and that we are entering a new age that is perhaps characterized by, among other things, a greater degree of great power competition and challenges to fundamental aspects of the U.S.-led international order that has operated since World War II.

Mr. FORBES. So I wouldn't be changing your words if I were to say that, based on the world situation, developments that have taken place within the last 12 to 18 months, that, in your view, that could suggest that the Navy should at least relook the direction that they are heading with this UCLASS program?

Mr. O'ROURKE. I can't make a recommendation, as you know. But what I can say is that, with other people out there—and I am not asking anybody to accept——

Mr. FORBES. I understand.

Mr. O'ROURKE [continuing]. My own judgment about whether we are entering into a new strategic era, but a number of other observers are saying that.

And it is enough to tee the issue up for the committee and the Congress as a whole to make its own decision as to whether we are entering that era and, if so, whether we should then have a reassessment of defense plans and programs.

Mr. FORBES. Okay.

Mr. Martinage, I would like to clarify something Mr. Courtney asked and—just to make sure that we heard you correctly.

When you talked about the 1,000-pound element, you were not talking about 1,000 gallons of fuel, I don't think. You were talking about a 1,000-pound payload.

But correct me if I'm wrong, I mean, because I could have heard it wrong. I just want to make sure that question——

Mr. MARTINAGE. That is correct.

Mr. FORBES. And, based on that, Mr. Courtney is exactly right. We have a lot of classified information on here. He is perfectly correct on that.

But it is your understanding, as I take it from your testimony, that you believe the requirements out here would relate to a 14-hour endurance requirement in the air?

Mr. MARTINAGE. That is correct. Yep.

Mr. FORBES. And a 1,000-pound payload max. Is that correct?

Mr. MARTINAGE. That is my understanding from what is in the public domain.

Mr. FORBES. Now, if that is accurate and if we have flexibility to add additional—would we have flexibility, in your opinion—if you have those two requirements or one of those two requirements, can we add additional payload capability to this platform?

Mr. MARTINAGE. Meeting the 14-hour unrefueled endurance requirement I think would likely preclude a significant increase in payload.

You could probably increase it some and trade off some endurance, for example, by putting fuel in the bomb bay or something like that or putting external weapons carriage and reducing survivability and reducing some endurance.

If I could, I would like to build on a comment that was asked about the evolving and adapting over time, if——

Mr. FORBES. Please.

Mr. MARTINAGE. I think that is important for this aircraft design to be able to evolve and adapt over time, but you need to get the shape and the propulsion path right or you are stuck forever in terms of the payload and the survivability. You can't undo those things.

You can. You can build a new jet. But if you get the shape wrong at the start and you get the propulsion path wrong at the start, you really can't go revisit those things and add payload and survivability later.

So driven by that 14-hour unrefueled endurance choice, choices are being made on the shape and the propulsion path that really can't be undone, and they will lead to an evolutionary dead end for the aircraft.

And that is why I am concerned. I think having a spiral development approach is a good one and you could do that with a balanced design.

So you could take that aircraft that ultimately has 8 to 10 hours of endurance, 3- to 4,000 pounds of payload, a very low signature, and 24 to 48 hours of refueled endurance, and you don't have to get there all at once.

You could field the basic shape and propulsion path. Over time you could add more advanced edges and coatings to get the survivability. Over time you could add additional weapons that it can carry. Over time you could add sensors to it.

But, again, it is fundamental to get the shape and the propulsion path right at the start. And right now that is being driven by the 14-hour requirement, and you have to ask yourself why 14 hours.

Mr. FORBES. Now, let me ask you about that.

Assuming your testimony to be correct and we have a 14-hour endurance figure and we have the 1,000 payload max, what does that limit me from using in terms of payload?

Mr. MARTINAGE. Well, the devil is in the detail, sir, and I don't know yet because we haven't seen the designs.

But one is you won't be able to carry enough weapons to saturate—say you were going after a surface combatant with short-range air defenses like a Luyang II or Luyang III in the case of the PLA [People's Liberation Army] Navy.

You would want a lot of individual weapons to saturate their point-defense and then to take out, to neutralize the ship. Four SDBs [small diameter bombs] is never ever going to do it for you.

I think it is instructive that we don't think about 1,000 pounds of payload under the F–35 or the F–18. Why would you think about it on UCLASS?

The other thing is what types of weapons. And that gets to not the weapons carriage, you know, how much it can carry, but, rather, the volume of the bomb bay. And that is just unclear. That is something that needs to be looked at.

But in order to go after some of these types of targets that are defended targets, you would want some standoff capability. And it is unclear, as I read what is available, whether or not those types of weapons will fit in the bomb bay of this current design.

Mr. FORBES. Now, if I can take the advocate's role for a moment and if we are already planning to procure 80 to 100 long-range bombers, is there a need for a UCLASS program focused on the strike mission?

Mr. MARTINAGE. I would say, (a), we are a joint force and it is good to have multiple options for the Commander in Chief to pursue.

Aircraft carriers don't need basing, and they can respond quickly to crises wherever they are without having to ask for permission for basing and access. They complicate an adversary's defensive challenges because you can come from multiple directions that they might not anticipate.

And then I would ask back: If the carrier doesn't have a long-range strike capability, what is the point? It is supposed to be the major power protection arm of the U.S. Navy. If we can't project power where and when necessary, why do we continue to invest in it?

Mr. FORBES. Mr. Brimley, if I could ask you to, one, describe for the committee, if you would, cost imposition strategies.

And if you could give us your thought as to whether or not pursuing the current course, as you understand it to be for the Navy for UCLASS, would impose any cost imposition on any of our competitors.

And if we pursued another course, would that have any cost imposition aspects to it?

Mr. BRIMLEY. Thank you, Mr. Chairman.

I think, for me, the most instructive case is just to look at what is going on in the Western Pacific. I mean, since the late 1990s, I think China and the PLA have gone down a path of imposing costs on us.

Well, what does that mean? I think they have spent the preponderance of their defense budget over more than a decade or more—you know, close to two decades now, trying to literally push us farther away. You just have to look at their development of increasingly long-range and precise anti-ship ballistic missile technology and, also, anti-ship cruise missile technology.

This is the fundamental problem, as my colleague Mr. Martinage talked about. If we can't project—we don't have a capability from the aircraft carrier to project much farther than 1,000 nautical miles, then it is very hard for us to be able to deter behavior.

And I think that, were we to invest in a platform that could actually project power over that distance or longer, you could significantly complicate our adversaries' calculus and impose costs on them, as well.

The other way I tend to think about this, too, is, you know, our allies and partners in the region are looking to us to be able to be with them in moments of crisis and moments of tension, and if we have a threat that forces the power-projection hub, the central power-projection platform of the U.S. military, to stay well, well outside the first island chain, for instance, it could really undermine U.S. foreign policy and national security strategy with our allies and partners.

Mr. FORBES. A more ambitious UCAV program would likely cost more.

Can you tell me a little bit about your opinion of what the value proposition would be for this?

Mr. BRIMLEY. Mr. Chairman, yes. You know, I can't give you specific cost estimates. I am sure the Navy and elements of the Pentagon could give you that.

But I would just suggest that going down what I called in my testimony a strategic UCLASS cul-de-sac—I mean, if we invest all this money and all this time to get a system that provides perhaps some better maritime demand awareness around the carrier strike group, but doesn't buy down any sort of risk regarding our ability to project power, then, in my mind, it is a waste of money and it is a waste of time.

So even if—I mean, let's just assume for a moment that a more ambitious UCAV could cost, say, 20, 25, 35 percent more than the equivalent number of systems of a less capable, less mission-centric capability that can't project power. To me, as a civilian, as a policymaker, that is a trade worth making. That is an investment worth making, in my view.

Mr. FORBES. Mr. McGrath, can you put the UCLASS program in an historic context, as you look at it, and maybe tell us how you see it fitting into a broader U.S. defense strategy both in Asia and more broadly.

Mr. McGRATH. I will take the second part of the question first and maybe an answer to the first part will reveal itself.

Our geography is not going to change appreciably in the near future. This Nation depends on its Navy and its Marine Corps to a large degree for much of peacetime-shaping and presence missions and transition-to-war duties around the world where our far-flung interests are.

That capability is something no other nation on Earth has, and it is a capability that a nation with our geographic constraints has to have if we wish to be influential, thousands of miles from our shore.

Because of the nature of the threat and the obvious desire of strategists around the world to try and keep naval forces from being able to generate significant power near their shores, because they are trying to keep the carrier away—further away, as Mr. Brimley was talking about, we have to counter that.

If we do not counter that in a way that continues the relevance of the aircraft carrier as our Nation's primary power-projection platform, we either have to acknowledge the end of American naval dominance or we have to figure out some way to replace that power projection.

I don't know what that is. I don't know—I don't know another platform or series of platforms or ensemble of platforms in the Navy that could—for the amount of time that an aircraft carrier can generate power that could match it. It is one of the reasons we build them and operate them, is because they are very efficient producers of combat power.

I don't have a good answer for you on the first part of your question.

Mr. FORBES. Mr. Langevin, like Mr. Courtney, has spent a lot of time looking at this issue and other naval issues, and we would like to recognize him now for 5 minutes

Mr. LANGEVIN. Thank you, Mr. Chairman. I want to thank our witnesses for being here today, and I apologize with the votes and all that. I may have, some of my questions, I hope they are not redundant, but I need to have these answers, and I would appreciate any insight you could give us. Again, thank you for your testimony.

Mr. Chairman, I appreciate the committee's attention to this very important program, and I thank you for the attention that you have given to this, and I, again, thank our witnesses for their insights in appearing before us this afternoon. As you highlighted in your testimony, this is a debate not about a program really but about the future of carrier-based aviation. The overall pattern of unmanned systems has been initially the description, the substitution of such systems for the three Ds—the dull, the dirty, and dangerous—if you will. And certainly persistent ISR is dull, though vitally important, which is why, of course, we have Global Hawks and Reapers and Predators, Fire Scouts, and other systems focused on that ISR mission. So my concern is that unless the Navy asks industry for the right capabilities, we could preclude right at the outset UCLASS's ability to stand in for manned aircraft for future dangerous missions such as ISR, denied environments, or initial strikes to take down integrated air defense systems and heavily defended targets.

We are here today to make sure what could be a truly revolutionary capability for the future air wing achieves its full potential. So my first question is, could you talk to the subcommittee, could you talk us through the design tradeoffs that will be necessary should the current unrefueled persistence requirement stay as is, and what options would be available to air vehicle designers if it was lowered? I will start right down the line.

Mr. O'ROURKE. I will just make a general comment that any one platform exists within an envelope of tradeoffs and that there are certain characteristics that can be achieved only to a certain degree in the presence of other characteristics, so range and endurance would be one. Payload would be another. Stealth and survivability would be a third. And cost would be one. And that and maybe one or two other attributes would establish a zone or an envelope within which you would make these kinds of tradeoffs. I am going to stop right there and let the other witnesses answer it in more detail.

Mr. LANGEVIN. Thank you.

Mr. MARTINAGE. I like to think about the design trade for carrier-based UCAV to really be driven by one—in all cases, it has to be able to operate off the carrier, which constrains its size.

But beyond that, think of a triangle, where you have unrefueled endurance on one corner, payload, mission payload, including strike payload, as another corner, and then survivability as a third corner. Anything you do to any one of those affects the other two. So when you say 14 hours of unrefueled endurance as one of those three parameters, you necessarily have to reduce what you might otherwise do in terms of survivability, and mission and strike payload. So the implication, to directly answer your question, sir, is the 14 hours of unrefueled endurance forces reductions or increases signature or reduces stealth and reduces payload in all its forms, including volume.

If you relaxed that 14 hours of unrefueled endurance, you could significantly improve stealth, which would get us into a classified conversation which we can't go into, but there is a lot more you can do there. And you could probably triple or quadruple the payload. And that payload doesn't have to be all used at once. That can also be your growth for the future in terms of size, weight, power, and cooling for new mission systems, new sensors, new weapons that you might want to integrate into the airframe in the future. But if you don't have that margin built in, in terms of mission payload capacity, you can't grow the aircraft in the future. I hope that answers your question.

Mr. LANGEVIN. Thank you.

Mr. BRIMLEY. I would just add, Congressman, I very much agree with my colleagues' statements. I would just say the number one operational challenge facing force planners, defense planners, the Commander in Chief, is do I have the option to penetrate an adversary's anti-access/area-denial network and hold that risk, their capabilities. And so as a civilian policy analyst, that is the number one operational challenge that I would ask the Navy to prioritize, and I think if you do that, you prioritize strike capacity, stealth, payload. Probably the last priority is really unrefueled endurance because that forces you to make all sorts of other compromises.

So I would ask, perhaps in your next panel with our Navy colleagues, you know, what kinds of design benefits could there be if you prioritize the strike side, the strike and stealth aspects of this design? I think that would open up all sorts of other possibilities that, frankly, would give better options to civilian leaders if we were actually to engage in some sort of conflict, or at the very least pose a more credible deterrent capacity overseas.

Mr. LANGEVIN. Thank you.

Mr. McGRATH. Mr. Langevin, I cannot improve upon those answers.

Mr. LANGEVIN. Thank you. My time is expired. I do have other questions, but I guess I should submit those for the record, Mr. Chairman?

Mr. FORBES. If you don't mind, Jim, we are going to submit a group of them for the record, and we would love to have your questions in there.

Mr. LANGEVIN. Very good, Chairman.

And I thank our witnesses for their insight and testimony.

And I yield back.

Mr. FORBES. Mr. Courtney, did you have——

Mr. COURTNEY. Just really quickly. One of the tradeoffs if you give up unrefueled persistence and reduce the thousand pounds to a lower weight, you are also creating another sort of challenge, aren't you, in terms of needing to have a refueling capability, which I guess the question is are we good to go as far as having that for unmanned air systems? I mean, that sounds like a whole new set of challenges, isn't it, in terms of making that work?

Mr. MARTINAGE. Well, the first thing I would say is 8 to 10 hours of endurance is still a long combat radius, so somewhere around 1,700 to 2,000 nautical miles, which is roughly triple every other aircraft on the carrier deck. So that is a pretty big operational reach improvement for the carrier air wing.

In terms of unmanned air-to-air refueling, that has been demonstrated already by surrogate aircraft as part of the UCAS–D program. Air Vehicle II is plumbed to do air-to-air refueling. It was originally part of the UCAS demonstration program. I don't know if the Navy is going to go forward with that, but I would say the technology is pretty mature and has been demonstrated with surrogate aircraft already.

Mr. MCGRATH. Sir, there was an open source report earlier this week, and I have it referenced in my written testimony—I don't remember exactly where it was—but where the Northrop Grumman program manager for UCAS says that they are prepared to demonstrate it during summer testing on board USS *Theodore Roosevelt,* and the UCAS Navy program manager indicated that if—excuse me. The contract has a provision in it, a clause in it, to do this. Northrop Grumman says they are ready to show it. The Navy program manager indicated that they were looking at trying to get to that, so we will have answers to these questions probably this summer, this coming summer.

Mr. FORBES. And if I could just end with one question, kind of a follow-up on what Mr. Courtney's is, I don't think there is a question that we have the capability to refuel, but the problem that I think Mr. Courtney may be getting at, too, is you would need a tanker or something to do it, which means we would have to have another asset there to do that, which is at least a question for us to raise.

But if you look at this triangle that, Mr. Martinage, you correctly raise between endurance and payload and stealth, if we had to pick a priority between those three things, what would the priority be, one? And, number two, if we continue in the direction that the Navy is heading, which is primarily an ISR asset, does that provide any additional capability than what the Navy already can do? And I would love to hear your responses on any of the four of you if you could address either of those two questions.

Mr. O'Rourke is looking at you, so I think he is——

Mr. MARTINAGE. I will start. I will start. In terms of the triangle, sir, I would, frankly, push back and say I wouldn't pick any one of the three. I would pick all three. And I would pick a balance of the three that allow it to perform the operational mission that I mentioned at the beginning, which is projecting power from range from the carrier, ISR and strike, in anti-access and area-denial——

Mr. FORBES. So then maybe if I could rephrase my question, not to compound it but to get an answer, you would say that the number one mission that you believe the UCLASS needs to be doing is the projection of that power through A2/AD defenses. Is that fair?

Mr. MARTINAGE. That is fair, and I would say the way I would prioritize it is that first thing, the aircraft needs to be able to go find and hit a target without being shot down. So what that means fundamentally is you need a certain level of stealth and the right weapon to hit the target and the right sensor to find the target. That is one.

Then, two, I want to have enough operational radius from a tanker or a carrier that I can do that from some reasonable range. I think 1,000 miles would be a good figure to put down. And then, third, I would want as many weapons and as many different kinds of weapons as I could fit into that aircraft. So that would be my hierarchy: find and kill targets without being shot down, do it at range, and have has many weapons in the magazine and different kinds of weapons as possible.

To answer your second question, sir, about what additional would this type of aircraft provide, it would provide the carrier strike group commander a long-range, long-persistence ISR asset for maintaining maritime domain awareness around the carrier battle group and then potentially finding targets for the manned air wing.

Mr. FORBES. That they could not do today?

Mr. MARTINAGE. No. No. I think that the MQ4C Triton, which is land-based, could do that mission, but that carrier strike group commander would tell you that is not organic to me. I can't control it, so I want something that I can control.

Mr. FORBES. That is a fair statement.

Mr. MARTINAGE. And then the other option that the Navy could pursue is things like MQ–8B/C Fire Scout off of any air-capable ship, so including all the destroyers and so forth; that has a potential of 8 to 12 hours endurance, and it could do maritime domain awareness around the carrier battle group that way. To me, that would be a much more effective and affordable way to get that ISR, rather than dedicating what really should be an integral part of the carrier air wing in the future for ISR and strike.

Mr. FORBES. Good.

Mr. Brimley.

Mr. BRIMLEY. Sir, I would just step back for a second, and if I could just make a broader strategic comment. Of course, the primary decision calculus here and for this committee and for the Navy and for the Pentagon ought to be, how can we make sure that the carrier strike group is relevant in the conflicts of the future?

But I think the broader strategic dynamic is concerning. We are only at the beginning stages of a revolution in unmanned, in increasingly autonomous systems. More than 75 nations are now investing in this kind of technology. And the proliferation dynamic is happening, and it is happening quick. As an analyst, what concerns me is if we spend 5, 6, 7 years walking down this path of making marginal improvements via an unmanned system to organic ISR for the carrier, that is time and money we are not spending in maintaining our military technical dominance and advantage in these early stages of what I think will be a very disruptive shift in the global military balance of power.

So I see this as a window of opportunity, and it is a finite window of opportunity. I am not a technical expert, but it strikes me that we ought to make sure that we spend the time necessary to get these requirements right before we start walking down a path that will close doors for us potentially and potentially undermine our military technical power.

Mr. McGRATH. Chairman Forbes, any strike group commander would love to have more ISR around his or her strike group. I mean, it is security. It is safety. It is knowing what is around you. The question is, do we have an adequate supply of platforms to provide that today? Especially when you have zero platforms today that can penetrate at the ranges we are talking about that make them operationally relevant in a contested A2/AD environment. This argument that the Triton and the P–8 aren't organic is interesting, but neither is most of the fuel that strike group commander is going to use in a campaign. It comes from the Air Force and tankers that fly and are in tanker tracks. I mean, there is some tanking on board the carrier, but a good bit of that campaign-level tanking will come from somewhere else. So I am not sure why the ISR coming from somewhere else is that big of an issue. We need the strike.

Mr. FORBES. Gentlemen, thank you so much for sharing your thoughts with us, helping us to create the questions we need to be asking and those answers, and we appreciate your time. Again, as we have apologized to you for these votes that delayed us, but we look forward to picking your brain in the future. And as Mr. Langevin pointed out, we have a number of Members who have some questions they will be submitting to you if you don't mind submitting those back for the record.

And with that, this panel will be—do any of you have any comments that you would like to briefly offer for the record?

Mr. McGrath.

Mr. McGRATH. I would just like to deal with one of the counterarguments, and I hear this from very serious people. And that is if you want another acquisition nightmare—and they compare it to, you know, you name it—go down the path that you are headed in, that we want to go in. And what bothers me about that is there is this implicit sense that we cannot do hard things well anymore. I think that is just not true. I think hard things are hard, and we cannot rush to a mediocre set of requirements out of fear that we can't do better. And I think this committee's leadership on this and continued pressure to try to ensure that we do better is required.

Mr. FORBES. Thank you.

Mr. O'Rourke.

Mr. O'ROURKE. Just two quick comments. Returning to Representative Courtney's earlier question about refueling, the solution that we reach, whatever it is, is something that I do think we need to consider in relation to the current refueling burden on the carrier. The carrier's own organic refueling ability does influence the range at which it can operate, especially in situations where it might be far away from land-based refueling. And I don't know what the impact is of various decisions on UCLASS in terms of the net burden on the carrier's organic refueling ability, but more than one of the people that briefed me from the Navy and industry did bring this up. So that is, I think, a factor that needs to be considered in the overall situation. And that's neutral as to the outcome of UCLASS, but from a system of systems point of view, what is the net impact on the refueling situation of the carrier, which influences its combat radius.

And the second, Mr. Chairman, goes to your question about the triangle and where we should be inside of it. I tend to think of it as a square rather than a triangle, because I think cost is the fourth factor. If you freeze the money that is available at a certain point and take it out, then you are possibly constraining the ability to imagine what the plane might be or what you might be able to achieve. So I see the trade space as having four corners, the fourth corner being a variable relating to the funding that is available for the program. And in terms of where you wind up inside that trade space, it is my hope that the six questions that I outlined at the outset will help people to think that issue through.

Mr. FORBES. Good. I am sorry.

Mr. Martinage.

Mr. MARTINAGE. Mr. Chairman, I would just like to follow up on a point Mr. McGrath made, and that is a point that has sort of been implicit to a lot of this conversation, is that a more balanced, more capable UCAV is somehow going to be more expensive and higher risk. I would just like to push back on that. When it comes to stealth/low-observables, it is more a choice about shape and propulsion path than it is about cost. Yes, there are marginal costs with the edges and coatings and sensor integration, but they are marginal. The big choice is where you go with the shape and the propulsion path of the aircraft. And that affects aerodynamic performance and other things, but it is not so much a cost driver. You tend to pay for these aircraft by the pound, regardless.

And the second thing is the technologies to achieve a balanced design, from low-observables and stealth to the payload capacities we have talked about to the combat radiuses that we have talked about, are all low risk. They have all been demonstrated. So if others come in here and say, Well, this type of LO [low-observables], or this type of LO across this frequency spectrum hasn't been demonstrated, it is just not true.

Mr. FORBES. And, Mr. Brimley, we will let you have the last word.

Mr. BRIMLEY. Well, sir, I guess on behalf of my colleagues, thank you for holding this hearing. I think there are—at least most of us I am sure agree, this is probably one of the top three to five defense design strategy procurement issues that are facing the Nation and the joint force writ large. Thank you for identifying this and zooming down on it.

One final point, as I think most of us outlined in our written testimony, the need to be able to pose operational challenges, to be able to penetrate anti-access and area-denial networks, is something that has been consistently enumerated since at least the 2006 Quadrennial Defense Review. So it is not as though this requirements debate is new. It is not as though this challenge facing the Nation and facing the joint force is in any way new. So I think, at least from my perspective, I find myself being able to draw on a pretty rich history in terms of the need to be able to set design requirements to be able to posture the joint force to be able to succeed in the future security environment.

And finally, just to echo what my colleague just said, you know, that plane that landed on the aircraft carrier last year was a certain shape and a certain design specifically because in the very beginning of this program, those kinds of design features were prioritized. And so I think, even from a joint force-Navy perspective, not so long ago, the three of us and our views, that was the view of the Pentagon. That was the view of the U.S. Navy. So I don't think that we are advocating sort of a technological fantasy or some kind of argument that isn't well within the mainstream of where force designers and planners have been for a long period of time.

Mr. FORBES. Gentlemen, thank you so much again for your time.

And with that, we will recess for the next panel.

[Recess.]

Mr. FORBES. Gentlemen, if you would have a seat.

I am sorry to detain you for these votes, but thank you for your patience. We have already given our opening statements earlier. We won't bore you with having to listen to those again, so we are going to go right on to your opening comments, any that you would like to provide, and I take it we will go in the order that you are seated, so with that, Admiral, we will turn it over to you.

STATEMENT OF VADM PAUL A. GROSKLAGS, USN, PRINCIPAL MILITARY DEPUTY, ASSISTANT SECRETARY OF THE NAVY FOR RESEARCH, DEVELOPMENT AND ACQUISITIONS, DEPARTMENT OF DEFENSE

Admiral GROSKLAGS. Thank you Chairman Forbes, Representative Courtney. Thank you for the opportunity to appear before you today to talk about the Navy's UCLASS program.

I need to start just by saying that the Navy is fully committed to the development and rapid fielding of an affordable persistent intelligence, surveillance, reconnaissance, and targeting, or IRS&T system, with a precision strike capability. The system will be based on an air vehicle design which ensures that we meet our threshold capabilities while being optimized to enable future mission capability enhancements, particularly in the areas of sensor payload modularity, weapons payload, and mission effectiveness, sometimes put under the moniker of survivability. The UCLASS key performance parameters and key system attributes, as defined and documented in the Service Approved Capabilities Development Document, remain consistent and stable.

That document was signed over a year ago by the Chief of Naval Operations and has not changed. The accompanying UCLASS acquisition strategy requires offers to be compliant with all the threshold requirements defined in a CDD [Capability Development Document] at a not-to-exceed cost per orbit, while incentivizing industry to propose systems and solutions that enable those future improvements and enhancements. More specifically, those proposed air vehicle solutions will be required to show basic design parameters that support the future growth, designs that can be affordably modified and enhanced over time to meet the future multi-mission needs of both the Navy and the joint force without major aircraft redesign.

We are on a path to achieve this growth capability without sacrificing the affordable near-term persistent ISR capability. We have had 4 years of very close engagement with industry, including technology maturation contracts which have culminated in the recently completed preliminary design reviews for four candidate solutions. This close engagement has provided the Navy with significant insight into industry capabilities which results in our confidence that affordable, technically compliant UCLASS design solutions are achievable within the targeted timeline and which take into account the plan form, the air vehicle plan form shape, and propulsion path characteristics that are needed to ensure that we can grow to the capabilities mentioned earlier.

It is also important to note that UCLASS will be a complementary and enhancing part of our carrier strike group. As part of the air wing, it supports the joint force and the Navy across our full range of military operations. UCLASS will make our carrier strike groups more lethal, more effective, and more survivable. I looked forward to your questions.

[The joint prepared statement of Admiral Grosklags, General Guastella, and Mr. Andress can be found in the Appendix on page 101.]

Mr. FORBES. Admiral, thank you.

Mr. Andress.

Mr. ANDRESS. I am going to let General Guastella go next if that is all right, sir.

Mr. FORBES. I think that would be a wise decision. General.

STATEMENT OF BRIG GEN JOSEPH T. GUASTELLA, USAF, DEPUTY DIRECTOR FOR REQUIREMENTS (J–8), JOINT STAFF

General GUASTELLA. Chairman and Representative Courtney, I appreciate the opportunity to come and testify today as well. I work for the Vice Chairman. I facilitate the JROC [Joint Requirements Oversight Council] process. And as many of you know, the JROC establishes requirements for our warfighting needs. The UCLASS requirements have been established by our most senior warfighters. They are the individuals that are responsible to organize, train and equip, not just their services but the entire joint force. All the services are represented at the JROC.

They established the UCLASS requirements not by looking through the lens of just the UCLASS system, but they evaluated the entire joint portfolio of ISR and strike assets to set these requirements, and certainly the JROC highly values carrier-based or sea-based ISR and strike platforms.

So I would like to read a sentence or two from the JROCM [Joint Requirements Oversight Council Memorandum], written on the 19th of December of 2012: And the requirement for UCLASS is for an affordable, adaptable platform that supports missions ranging from permissive counterterrorism operations to missions in low-end contested environments to providing enabling capabilities for high-end denied operations as well as supporting organic naval missions. Essentially that requirement strikes a balance between affordability and performance.

And like I mentioned earlier, you cannot look at the UCLASS requirement through a single lens, but only as you look at it how does it fit into a joint portfolio of assets, from permissive air-breathing to more advanced air-breathing assets to include space-based assets. Where is its role in there? And that is how those requirements were derived. While this is an open hearing, sir, if you need us to come by later and discuss some of the other systems or performance parameters, we are happy to do that. Sir, pending your questions.

[The joint prepared statement of General Guastella, Admiral Grosklags, and Mr. Andress can be found in the Appendix on page 101.]

Mr. FORBES. General, thank you.

Mr. Andress.

STATEMENT OF MARK D. ANDRESS, ASSISTANT DEPUTY CHIEF OF NAVAL OPERATIONS FOR INFORMATION DOMINANCE

Mr. ANDRESS. Yes, sir. Thank you for the invitation to speak. As the resource and requirements sponsor for UCLASS, it is very satisfying to be here and see such strong advocacy for one of our programs. I would say it is even beyond just the advocacy for unmanned sea-based ISR and strike. It is a desire for more. It is a desire for more weapons payload. It is a desire for more unrefueled or lower unrefueled range. It is a desire to add non-organic tanking to those requirements, and it is a desire for greater stealth. All these are great capabilities that we have captured and assessed through our requirements review process over the last 4 years.

I am perhaps more appreciative of the subcommittee's desire to look for answers that balance this demand for increased growth on a program, which we can talk about tradeoffs in performance and capabilities, but those tradeoffs in performance and capabilities in and of themselves have to be balanced by cost, schedule, and technical risk. I believe the answer that we are seeking for is that the Navy, working with the Joint Staff, have balanced not only the capabilities tradeoff desired, but balanced these capabilities against the cost, schedule, and technical risk we need to succeed.

As I go into the rest of my opening testimony, I want to highlight that I am going to only be able in an open hearing to talk to you about the process we use as we go through requirements building, and I will be happy to come back and talk to you more about specific threats, sensors, and others that are rolled into the requirements capabilities.

Mr. FORBES. Mr. Andress, could I ask if you could suspend your opening remarks for just 1 minute.

Mr. Courtney, do you have any questions that——

Mr. COURTNEY. I just found out my appointment just cancelled.

Mr. FORBES. I am sorry. Mr. Courtney was going to have to leave, and I wanted to make sure he got his questions.

Mr. Andress, I am sorry to interrupt you. Please go ahead.

Mr. ANDRESS. No problem. So, as General Guastella has pointed out, I am going to go into the requirements and how we float the requirements from start to finish. I am going to talk to you about the missions that those are trying to achieve, and I am happy to follow up with more specifics on the threats and the parameters that we are trying to target with UCLASS.

So UCLASS will be permissive ISR and strike in the near term as we have prioritized getting this initial capability to the fleet and operating off a carrier in a technically viable, timely, and affordable manner. But the overall system must be able to operate in a contested environment and support high-end operations in the 2020s and beyond to pace the threats we believe will be present as part of the larger carrier air wing.

I want to make sure that you understand the depth of requirements analysis that has gone into this very high-level requirements statement. The fundamental gap that the program is bringing came from a capabilities-based assessment looking across not only the future carrier air wing of 2025 and beyond, but we also looked at the carrier air wing operating in a joint environment. The gap that was most prevalent from running multiple scenarios is persistent ISR and strike from the sea base. That capabilities-based assessment led to what we call an initial capabilities document. This document is endorsed by the Joint Staff, and it specified a range of tactical scenarios in which we will need this program to operate. These scenarios have been reviewed and endorsed repeatedly over the last 3 years. They have not been static. They have been revisited not only from the standpoint of the threat scenario. They have been revisited with the Intelligence Community to look at what is the evolving threats that it must face? What are the ranges? What are the frequency bans? What are the distance we must be at? So that has not been a static process. It has been a recurring process leading up to this RFP.

The scenarios, and we call them design reference missions, begin with permissive ISR and strike and then move into contested ISR and strike against littoral threats. Think of this as small boat threats to naval forces. Next is contested ISR and strike against coastal land-based threats. Think of this as coastal defense cruise missiles, other emerging threats. And finally, anti-surface warfare scenarios. Think war at sea against near-peer adversaries. These design reference missions also include the need to both give and receive aerial refueling.

These scenarios which are based on how the COCOMs [combatant commands] intend to fight and win in the next decade, drove the right balance of endurance, sensors, weapons, and self-protection for UCLASS as a member of a carrier strike group that in this timeframe will include Joint Strike Fighter, E2D, and Growlers equipped with next-gen [generation] jammer capabilities. The Navy made the decision to field these capabilities in increments primarily based on cost, schedule, and technical risk. Through extensive engagement with industry, we believe the incremental ap-

28

proach can be achieved while maintaining relevance against the threats to carrier air wing in 2025 and beyond. Thank you.

[The joint prepared statement of Mr. Andress, Admiral Grosklags, and General Guastella can be found in the Appendix on page 101.]

Mr. FORBES. Mr. Courtney, I am going to yield to you in case you have to leave. I appreciate you staying to ask your questions.

Mr. COURTNEY. Thank you, Mr. Chairman. I appreciate your courtesy again. That conflict evaporated. So I want to thank the witnesses for your testimony.

Admiral, I think I counted the words "modification," "enhancement" and "growth" probably about four or five times during your remarks and echoed in the other witnesses, and that, as you may have heard from the prior panel, I mean, was sort of a big focus of the discussion, which is just, you know, some of the witnesses were sort of posing the next step as almost this irrevocable decision. And, you know, I just wonder if you could just sort of maybe talk about that a little bit more about whether or not you think that the persistent non-refueling priority is something that is going to lock us into, you know, a system that can't sort of achieve sort of the goals of strike capacity that I think everybody agrees would be good for the country. So, again, I was wondering if you could comment on that?

Admiral GROSKLAGS. Yes, sir, and I will probably ask the requirements officer, Mr. Andress, to tag in as well.

And we have been very, very careful as we have built the request for proposals and flowed the requirements from the capabilities description document or development document down into a detailed specification to try and ensure that industry understands that we need a solution that can grow to future mission roles over time, should the Navy and the joint force decide to implement those. It is technically achievable. We have seen the designs that industry is likely to offer us through the preliminary design review [PDR] process. That report out of the PDR process that is required by NDAA [National Defense Authorization Act] language from a couple of years ago I believe you will see if not late this month, early in August. It does certify that those PDRs were complete. They were in accordance with the process. But the net result of that, from our perspective, was along with 3 years of very close observation of what industry can offer, we know that there are technical solutions out there that provide us the capability to grow to a more survivable—read low-observable—platform if we decide to go down that path.

There are things that we need industry to bring on day one in order to ensure that that is possible. We have seen that in their designs already, so we are comfortable that we can get there. We have also asked them to look at additional payload in terms of weapons, and we have asked them or required them to bring to us on day one additional capacity opportunity beyond, and I don't want to go into too much detail of source selection criteria, but additional capacity beyond the 1,000 pounds that was talked about earlier. The threshold requirement remains that 1,000 pounds on day one. That is our early operational capability. It is also important to point out that the aircraft is required to have two external

3,000-pound hard points which can carry fuel for refueling other aircraft. They can carry other weapons. They can carry other sensor pods for a variety of missions. We are ensuring that all of these capabilities or enhancements for growth capability are built in on day one. We don't intend today to implement all of those because we don't know where the requirements of the future will necessarily take this platform, but we want to ensure that it is not a dead-end solution for the carrier or for the joint force, that it is a very adaptable solution that can be incrementally grown in capability into the future, and we believe the requirements support that and our acquisition strategy that industry will see through the request for proposals reflects that as well.

Mr. ANDRESS. I could add to that from the requirements. We spent a lot of time in the analysis of alternatives looking through unrefueled persistence. We analyzed 8-hour. We analyzed 10-, 12-, 14-hour, and 24-hour endurance models. We viewed these, all the time we view these through those design reference missions. It is against the targets they have to kill, where the aircraft needs to be, how far it needs to be from the carrier, et cetera. And we assessed that against mission effectiveness, but we also looked at technical risk and cost. And so while I will need to be able to come talk to you about 8 and 14 hours from the mission effectiveness in a more classified setting, I can certainly talk to you in this setting about cost, which was a huge driver in the AOA, and some of the technical risks.

Twenty-four hours of endurance is, while the most cost-efficient, introduced unacceptable technical risk to one of our top performance criteria, which is carrier suitability. This is mostly driven I think by wingspan, payloads, and things like that. Eight hours endurance introduced by far the highest lifecycle cost of all the alternatives by a margin of over four to one. Fourteen hours, when you looked at the difference between 8 and 14 hours from a development standpoint and introduced negligible technical or cost risks, so the 14-hour requirement facilitated the optimal balance to achieve two 24 by 7 orbits at 600 nautical miles from the carrier or one orbit at 1,200 nautical miles from the carrier or a single strike mission—that is an orbit, persistent, 1,200 miles—or a single-strike mission at over 2,000 miles from the carrier. So we are very comfortable with the unrefueled requirement as it sits and that it doesn't limit our ability to grow to objective requirements across the other balances of weapons, survivability, et cetera.

Mr. COURTNEY. One other question. Again, obviously the House has acted with the Defense Authorization Bill for 2015 and, again, had the additional review that was included in the language. I mean, at this point the classified RFP is, I mean, that is imminent. Right? Is the game plan pretty soon in terms of when that is going out?

Admiral GROSKLAGS. Yes, sir. In fact, the Defense Acquisition Review Board that was planned to make the decision for the release of that document was scheduled for next Monday. The Deputy Secretary of Defense asked for a precursor meeting, and because several of the principals were out of the country this week, that meeting and the subsequent DAB were postponed until next week.

It is our expectation they will both happen next week, and the RFP release would follow that second meeting.

Mr. COURTNEY. I mean, at this point, if the law or the bill passes as written by the House—I mean, there is assessment requirement. I mean, is that something that you will just have to—I mean, obviously, it is the law. But I mean, it doesn't conflict necessarily with that RFP already having been released. Right? You are just going to have to comply with it as written.

Admiral GROSKLAGS. No, sir, it does not conflict with releasing the RFP.

Mr. COURTNEY. Thank you.

I yield back.

Mr. FORBES. Thank you, Mr. Courtney.

Admiral, I want to come back. Mr. Courtney was exactly right. We heard three words, ''modification,'' ''enhancement,'' and ''growth,'' by everybody, but there was another word we heard from everybody, too, which is ''affordability.'' Could you help me with the definition of affordability, because that seems like something you guys really want to get across, affordability? Between an aircraft carrier and an LCS [Littoral Combat Ship], which one is most affordable?

Admiral GROSKLAGS. I am not sure I can crisply answer that question.

Mr. FORBES. And the reason you can't answer it is because it depends on the mission that you want to accomplish. Isn't that true?

Admiral GROSKLAGS. Absolutely, sir.

Mr. FORBES. So if you are looking at a UCLASS, we can get a very cheap UCLASS platform if we just reduce down the requirements. The key for us is making sure we get the right mission that we want the UCLASS to accomplish. Is that a fair statement?

Admiral GROSKLAGS. Yes, sir, I think it is important that we get the correct requirements upfront, and that is where this program started, with the correct requirements.

Mr. FORBES. Can you help me with this part, too? You indicated to us that this document was signed a year ago or longer. Is that fair?

Admiral GROSKLAGS. The capability development document, or CDD, was signed in April of 2013 by the CNO. It is a Navy document at this point.

Mr. FORBES. The reason I ask that is because we had testimony by several individuals in our previous panel, including Mr. O'Rourke, who we put a lot of credibility in as kind of the historian that we look to for the CRS [Congressional Research Service], who says that there have been world events that have developed in the last year, including China's actions, the Russians and Ukraine, that perhaps would give a pause to look and see if the strategic look that we had at this program a year ago might not have changed in the event of world events. Do you think Mr. O'Rourke was wrong in his analysis?

Admiral GROSKLAGS. Not having seen his entire analysis, I don't think he is incorrect in that we have an obligation, frankly, to continue to look at our requirements over time to see if they are, in fact, evolving or need to evolve to meet any new threats.

Mr. FORBES. Would it be fair of this committee or subcommittee if we were to simply say that we have had certain events that have taken place in the last 12 months that have dramatically changed some of the ways that we look at our strategic goals around the world?

Admiral GROSKLAGS. I would say that is correct. I would also comment that we have continued to look at the—while they have not changed—we continue to look at the requirements for the UCLASS program. And, again, one of the key aspects is that we are building in that ability to adapt this platform to missions of the future, regardless of what they may be as long as they fit in the earlier discussion that we had.

Mr. FORBES. And I want to address those. I want to also then come back to what you just said. You don't fault this subcommittee for saying that, based on those changes, since you have just said that you need to continually be looking at those requirements, you don't fault this subcommittee saying before we pour the concrete and start heading down this very expensive program, that we should perhaps measure twice and cut once to make sure we get it wrong—right, and to make sure that the Secretary of Defense has a second look at it?

Admiral GROSKLAGS. Sir, I will let the requirements folks talk to that specifically, but I believe we have measured at least twice. We continue to measure, and in fact, over the last weeks, the Joint Staff and the Navy have continued to look at these requirements and, frankly, from my perspective, continue to validate what we have.

Mr. FORBES. I am going to give you plenty of opportunity to re-spond to that. First of all, I want you to know, I don't want you to talk about anything that is classified, and we know we have got all of that in our logistics. But we have heard the previous panel talk about this 14-hour time period.

Mr. Andress, I think you just mentioned that, and you said there was no technical risk or cost difference if we went from 8 hours to 14 hours. Did I understand that? Or marginal, maybe your word was ''marginal.''

Mr. ANDRESS. No. You misunderstood. The cost risk of going from 14 to 8 is dramatic. It includes both cost from fuel, cost from fuel burdens for tankers. It includes costs for additional integra-tion. Now, remember, you are an unmanned system, so now you have to integrate the system that you will refuel with in the unbur-dened tankers. All those now have to map to those tankers.

Mr. FORBES. So the cost of having 8 hours versus 14 hours.

Mr. ANDRESS. Is enormous.

Mr. FORBES. Much more expensive.

Mr. ANDRESS. Dramatically. Like I said, it is more than four times.

Mr. FORBES. So that is why you would like to lock into the 14 hour——

Mr. ANDRESS. That is not the only reason, sir. Cost was one fac-tor, and we locked in the 14 hours at both threshold and objective. It doesn't need to change to achieve our contested high-end oper-ations that we get at the objective requirement. It does not need to be 8 hours refueled when combined with those other capabilities, which are weapons, sensor, sensor range detection, survivability, et cetera. So I highlighted lifecycle cost because it drove a lot of the analysis of alternatives as we look to where and when this thing needs to operate in the dependencies.

Mr. FORBES. I think that is fair. Just like our LCS and carrier example, it is much more expensive to operate a carrier. Since we now have 14 hours that we are talking about, and we can at least talk about that figure, would you also agree that if you do lock into that 14 hours, that you have significantly limited the amount of payload that you could expand to at a foreseeable time given the platform that you would have?

Mr. ANDRESS. No, sir. We are not limited by the payload, the growth in the platform as we go from threshold requirements, permissive to the contested and high end.

Mr. FORBES. So then you would say that when we talked about that 1,000-pound payload factor, you would say that is inaccurate, that you could go much higher than that 1,000 pounds and still have that 14-hour endurance requirement?

Mr. ANDRESS. Absolutely, sir, and that is specified in the threshold to objective requirements that Admiral Grosklags has testified we would be able to achieve.

Mr. FORBES. Ok, now let me ask you this. If you look at in light of what I at least understand to have senior level guidance about the growing need to project power despite A2/AD challenges in the past four QDRs and in the Presidentially approved Defense Strategic Guidance, why did the JROC change UCLASS requirements away from the A2/AD last fall?

Mr. ANDRESS. I am not aware that the JROC changed the requirements away from it.

Mr. FORBES. Were the requirements changed? I am sorry. I didn't mean to interrupt you.

General GUASTELLA. Sir, the requirements have grown over the 3 years. They have been looked at. Actually, the JROC, in terms of your question before about measuring how many times, the JROC has looked at the UCLASS requirements six times over the last 3 years. Sir, most recently, the 4th of February it was looked at again, so very recently, especially in light of current events.

And like I mentioned before, sir, and as you have said, the world is more dynamic now possibly than even when the platform was first envisioned. The budget pressure, however, is more acute than ever, and a trillion dollars over 10 years is what we face. And so I think that the JROC is aware of that fiscal reality and ensured that it has been able to make—been forced to make performance and tradeoffs——

Mr. FORBES. And, General, that is my point. You need to help us with what we need as what the mission should be because we have also had the Navy come over here and say they want to park 11 cruisers. And we said, We disagree with you. We have had the Navy come over here and say, We don't want to do another carrier. And we felt that was wrong. So when you tell me that we looked at different requirements and we looked at budget requirements, I am asking you the strategic requirements because, as I understand it, one of—the previous panel would just disagree with what you think the ultimate mission might be for the UCLASS. They think that one of the most compelling needs that we are going to have with this platform is to be an integrated part of our carrier wing that can penetrate A2/AD defenses. What I hear you saying is that it needs to be the sophisticated ISR capability.

And so as I looked at the senior level guidance, we have heard over and over again their need to project these same kinds of A2/AD defenses. And so I guess my question is, not just based on budget or fiscal restraints, but what is it that caused you to change the ultimate mission goal that you had, or did that change?

General GUASTELLA. Sir, I think it is best to say that no asset serves a single purpose. And almost every asset in DOD serves the joint fight.

Mr. FORBES. Fair. What is the primary purpose? Would you agree with me that I am making choices between a primary purpose of an ISR capability or of a platform that is capable of penetrating A2/AD defenses?

Mr. ANDRESS. I will take it. I want to go back to the questions about the strategic. And it is very important to note that UCLASS, the CDD was signed last year by the CNO, and we have looked at it again through the JROC just a few months ago. The implications are that we have a shift from this post-Cold War permissive only mentality and that UCLASS has missed that in its requirements. And what I want to assure the committee is that the design reference missions that I spoke of—spoke of, follow, first—follow the Defense Strategic Guidance, speak to both the permissive environment as a threshold capability but have the requirements and objective capability which they will grow to to get at the contested and supporting the high-end A2/AD environment. Those were thought through, and the specific locations where the threats, the surface-to-air missiles we must face, the enabling factors to A2/AD were factored into the mission performance that UCLASS needs to meet.

Strategically, your question was, where does UCLASS fit in on that? The capabilities-based assessment says that UCLASS provides ISR and strike at longer range from the carrier, ranges I just spoke to, two orbits at 600 nautical miles, one orbit at 1,200 nautical miles, and single strike missions at 2,000 nautical miles. That strategic mission hasn't changed. The requirements have not changed. Our path to get there is consistent and is balanced. The capabilities are balanced against cost, schedule, and technical risk.

Mr. FORBES. So, Mr. Andress, can you provide us the assurance that this RFP will create a platform that will meet those threshold objectives?

Mr. ANDRESS. Absolutely, yes, sir.

Mr. FORBES. What about the objective requirements?

Mr. ANDRESS. Yes, sir. I think your question is, will the RFP address threshold requirements, and will it enable a path to objective requirements. Is that your question?

Mr. FORBES. No. Will it meet the objective requirements?

Mr. ANDRESS. This RFP is not designed to meet the objective requirements. It is not a part of our acquisition strategy, not when you balance it against cost, schedule, and technical risk of carrier suitability.

34

Mr. FORBES. Now, let me just ask you a couple final questions. Is the Navy abandoning the precision landing system developed and successfully tested during the UCAS–D effort for the UCLASS program?

Admiral GROSKLAGS. Sir, our long-term plan is to utilize the Joint Precision Approach and Landing System [JPALS], which is a separate Navy-run, Navy-managed program of record. A couple of advantages in doing that: The first is the UCAS–D precision landing system is a very proprietary system that was designed specifically for this demo program. It would require significant modifications to enable it to operate long term in the carrier environment operationally.

Alternatively, with JPALS, what we get is a solution that we are not only going to use for UCLASS but also for the Joint Strike Fighter. A Joint Strike Fighter also requires a precision landing system, and JPALS fits that need for both the F–35Cs on board the aircraft carrier and the F–35Bs for the Marine Corps on board our large-deck amphibs [amphibious assault ships]. So that common program serves multiple platforms for the Navy.

Mr. FORBES. Last question I will have for you. Could the UCLASS air vehicle described in the draft RFP operate in the South China Sea against the Chinese SAG [Surface Action Group]?

Mr. ANDRESS. Our most stressing design reference mission dealt with SAG capabilities, and I am happy to come by and show you what that exactly entails, what that threat is, where that threat is located.

Mr. FORBES. And when you do, could you also talk about the Taiwan Strait and the Black Sea and how that would operate there if you don't mind?

Mr. ANDRESS. Yes, sir. I would be happy to do that.

Mr. FORBES. Mr. Courtney, did you have any follow-up questions?

Mr. COURTNEY. Just a quick follow-up.

Mr. Andress, you mentioned again sort of some of the, I don't know what the word is, constraints or just realities that you have to kind of deal with in this whole process, and you talked about technical risk being one of them. Again, the prior panel, we had a little colloquy regarding the issue of refueling. If you were going to lower the fuel amounts on board, that obviously that would create an immediate requirement to have tankers be able to do this. Again, that is one of those technical, I don't know if the word is "risk," but challenges. I mean, tankers aren't really doing that right now in terms of refueling unmanned systems. Isn't that correct?

Mr. ANDRESS. Not to my knowledge, no, sir.

Mr. COURTNEY. So, again, when people sort of talk about well, let's lower the fuel capacity and that way we can add more strike capability, and if you are going to do deep strikes, it almost makes refueling, you know, mandatory. I mean, that is a whole another set of challenges that you have to deal with. Right? If you were trying to jump to that system right away.

Mr. ANDRESS. Yes, sir. It would be if we were trying to jump. We would not necessarily—remember the sea base is always mobile, and it is able to gain accesses that land bases do not always give us the luxury to enjoy. So when we talk about a threshold capability of 2,000 miles for a single-strike mission, that is a significant strike unrefueled. So no fueling requirement, no tanking along the way. That is significant. And in the threshold capabilities is where you get into where that strike mission—I mean, the objective capabilities that you run against those missions get into how much more difficult that strike mission would be. But at threshold on objective, we will have unrefueled tanking to support a 2,000 nautical mile strike.

Mr. COURTNEY. Right, pretty good range.

Did you want to comment, General?

General GUASTELLA. Sir, if I could, an absolutely valid question.

The JROC's approach is—if you look at it as a strike asset, it falls into a family of strike assets. And so, if there is very long-range targets, maybe that would be something more suited to different assets.

And so what we will do is tailor targets that are associated to the UCLASS's capability range to it and then assess other targets to different platforms.

And together, though, as a family of systems, it is how we feel we are best presenting a joint force for our country.

Mr. COURTNEY. I yield back.

Mr. FORBES. Well, General, in that regard, too, we have a family of systems for ISR as well and a lot of other capabilities that we are using for that.

And what we are looking at here is the Navy capability from a carrier platform because the big concern we have is the A2/AD defense is pushing our carriers out further and further.

So if we weren't worried about the Navy doing that, we could rely on our bombers or whatever might be involved in that family of assets.

Mr. Andress, just one last thing I want you to clarify again. And I think you have done this, but I want to make sure.

You said that cost was a huge driver in the analysis of alternatives. What alternatives did we take off the table because of cost?

Mr. ANDRESS. Sir, I spoke to the analysis of alternatives of different endurance ranges, from 8, 14, to 24. We took 24 off the table as we did the analysis both on mission cost and technical risk.

We then took off—as we went from the AOA, we saw that those two were still viable. Then you started to look at—from cost, it became the 8-hour mission as compared to mission effectiveness, and we went with the 13.6 hours.

Mr. FORBES. So you didn't take off additional payload or anything like that based on your cost analysis?

Mr. ANDRESS. Could I take that for the record to make sure?

Mr. FORBES. Sure. If you could just bring that over when we——

Mr. ANDRESS. It is a complicated mix of tradeoff analysis. I want to make sure I get you the right answer.

Mr. FORBES. And we can talk about that when we get into the classified setting.

Mr. ANDRESS. Absolutely.

Admiral GROSKLAGS. Sir——

Mr. FORBES. Oh, please. Please.

Admiral GROSKLAGS [continuing]. On that question, two other things that were taken off the table because they were looked at as part of that analysis of alternatives—and I believe they were mentioned earlier—was the Fire Scout—that was looked at as a potential solution for this mission requirement of ISR&T from the carrier—as was Triton.

And those were taken off for obviously different reasons, but the determination was neither one of them could meet this requirement for something that supported the carrier, wherever it happened to be around the world, real-time.

Mr. FORBES. Good.

I had promised all three of you that I would give you any time you needed for a wrap-up. And so I want to offer you that now.

And maybe, Admiral, if we could start with you and——

Admiral GROSKLAGS. All right, sir. First, I want to clarify something that was mentioned today just so we are crystal-clear.

You had asked a question about 14 hours and, if we add additional ordnance or mission systems to the aircraft, was that 14 hours still achievable.

I think the answer to that is it depends. We will find out from our vendors once we get the actual proposals back whether—as we add additional weight to the aircraft, typically that will reduce their endurance capability. We will find out whether, with additional payload, they can still meet 14 hours.

But the specific requirement for the day-one capability—and, again, this is the near-term day-one capability—is 14 hours unrefueled with a single laser JDAM [Joint Direct Attack Munition]. That is the basic requirement to get out there, perform the ISR&T mission with a precision strike capability.

We acknowledge in this specification, as they bring more weapons to the table, just like any other aircraft, be that manned or unmanned, that the endurance, the range, the performance of the air vehicle will decline over time.

Mr. FORBES. But, Admiral, help me with that, then, because I— you know, I just want to make that clear. And thank you for clarifying that.

Because I thought it was your statement—not yours, but maybe Mr. Andress'—that you could add that additional payload and continue to have that 14 hours.

And you are not saying you couldn't do it. What you are saying is you will have to check to see if you could do it. Is that a fair statement?

Admiral GROSKLAGS. Yes, sir. I just wanted to make sure you were clear on that.

Mr. FORBES. No, because that does help me.

Because if, in fact, the previous panel was correct in saying that they didn't believe you could do that and if, in fact, when you come back, your conclusion is we could not increase that 1,000-pound payload and continue with the 14 hours, then it becomes very important for us to ask what are we taking off the table by not being able to use more than 1,000 pounds of payload.

And if I could just ask one more thing. I am going to let you respond all you want.

Because the previous panel would suggest that there are a lot of those targets that Mr. Andress was talking about that you cannot handle with 1,000 pounds of payload.

So now please clarify for me to make sure I am not missing any of those points.

Admiral GROSKLAGS. No, sir. You are accurate.

And as the mission roles of this aircraft develop over time, what we are trying to ensure is the operational commander, whether that be the carrier strike group commander or a COCOM or a joint force commander in the area of operations, has the flexibility to make that decision: Do I want this 14-hour aircraft with a precision strike capability, albeit somewhat limited, or am I willing to give up a little bit of that endurance for a particular mission where I want to carry more ordnance or I want to carry a different sensor package or I might want to take additional fuel on my external hard points?

Those decisions, long term, we want to leave to the operational commanders. So we are trying to build an air vehicle that is adaptable to that situation.

I mean, I heard some earlier comments about an 8- to 10-hour aircraft that could carry 4,000 pounds and could not be seen by anybody. And, frankly, that doesn't exist. It does not exist, and it is not technically achievable today.

We have looked at the tradeoffs very, very carefully between all the mission roles, all the mission capability, weapons, low observability, endurance, cost, manpower to sustain: How does it fit on the carrier? How does it blend in with the rest of the carrier strike group? How does it blend in with the joint forces, as the general mentioned? And the 14 hours for that particular mission set at EOC [early operational capability] was really the sweet spot. We melded all those things together.

Mr. FORBES. And, Admiral, as you come back to us—and I don't expect you to have the answers today—it is night and day difference between what you said and what I maybe misunderstood from Mr. Andress, in that, if we are, in fact, saying that, if we lock into a requirement of 14 hours, we don't know for sure if we can increase this 1,000-pound payload—I am just—I am sorry. Go ahead.

Admiral GROSKLAGS. Yeah. I just want to be clear.

When we write requirements for any aircraft or, frankly, any system, we will have a baseline requirement that we can test them to. Okay?

So the baseline requirement for the fielding of this aircraft is to be able to comply with that 14-hour endurance requirement, the orbits that Mr. Andress talked about with a defined payload.

If we or the operational commander in the future chooses to change that mission payload, then we don't hold industry to provide us an aircraft that continues to meet all the other parameters that are encompassed in that baseline capability. We have to have something to measure them against initially, and that is what that 14 hours is.

Mr. FORBES. And I understand that, and I appreciate that.

I guess what I am trying to also say is that what the previous panel, I thought, was indicating to us is it might be more important for you to be able to have that increased payload as opposed to that endurance if, in fact, your goal is penetrating those A2/AD defenses.

And one of the things they would at least suggest is that you can't just come back and modify this platform quite as easy as it is being represented to be able to do.

In other words, it is not some modular thing that a commander in the field just says, "Okay. Today I would rather have 8 hours." You see what I am saying?

Admiral GROSKLAGS. Yes, sir. It is not that simple.

We know that when—based on the design solutions we have seen to date, we know that industry will be able to bring us something that has a greater weapons payload than what is in that initial requirement.

We also know and have been clear with the industry that, if they bring us that additional weapons payload, in that operational context, they will not be required to meet the 14-hour endurance requirement.

Mr. FORBES. And, if I could—again, I appreciate your clarification—that is the essence of what we have been saying.

If we lock into a requirement on this 14-hour provision because of affordability, because of whatever we are looking at, we just want to make certain that we are not going down a route that is going to take other options off the table which could be incredibly important, if, in fact, you view this UCLASS as important not just for ISR, but for penetrating an A2/AD defense.

And if you also say, for it to do that successfully, it has got to carry more than a 1,000-pound payload, then we are starting off in the wrong place at the beginning. And that is just what our concern is, and maybe that is a discussion we have to have in a classified session.

Admiral GROSKLAGS. I think part of it would be easier in a classified setting.

However, as we discussed earlier, we have seen through the preliminary design reviews that industry has design solutions that cannot only enable us to grow to a more mission-effective capability in terms of low observability without major modifications to the airframe that we are going to see on day one, we also have seen the capability to carry additional weapons in those same designs.

So, again, the 14-hour requirement is for that early operational capability single mission. That does not constrain us specifically from adding additional weapons capability to the aircraft. And, in fact, we have seen designs that carry considerably more than that 1,000 pounds. But we will not hold the providers of that to 14 hours with those additional weapons.

So, again, it gives us the flexibility in the future to decide, "Yes. We want to carry more weapons on a particular mission." We have to go through, you know, the process for certification for those weapons. So it will take a little bit of time, not day one. But it gives us that flexibility in the future to go there.

Mr. FORBES. Okay. And thank you, Admiral.

Mr. Andress.

Mr. ANDRESS. Yeah. I just wanted to build on that, and maybe I can offer some clarity.

Our threshold requirements for the 1,000 pounds and 14 hours and that has to be tested against, it doesn't change as we go from threshold to objective requirements. The requirements remain 14 hours endurance. The requirement for weapons grows beyond—it goes to greater than 1,000 pounds.

I think what Admiral Grosklags was pointing out that hasn't really been highlighted is, even at threshold requirements, this thing is going to have 1,000 pounds internal carriage and two hard points with 3,000 pounds.

So if the commander makes the decision to strap 3,000 pounds of weapons on the outside and the 1,000 pounds of internal carriage, you wouldn't expect it to still test to the 14-hour requirement. Does that make sense?

Mr. FORBES. Yes, sir. It does.

Mr. ANDRESS. I just wanted to make sure that was very clear.

Mr. FORBES. It does.

Mr. ANDRESS. And I think it will be very helpful for you when we show you the specific design reference missions at threshold and at objective that we need to achieve.

Mr. FORBES. I think it would. Thank you, Mr. Andress.

General, any?

General GUASTELLA. Sir, I appreciate the opportunity to testify today and realize it is an unclassified forum and it is difficult to present some of the analysis that JROC uses and how the UCLASS fits into that joint portfolio, but happy to come back if you have additional questions, sir.

Mr. FORBES. Good.

Well, gentlemen, thank you so much for your patience and for being here to help us. We thank you for your service to our country. And we look forward to getting together again in that classified session to be able to talk further about it.

And, with that, Mr. Courtney, if you don't have anything else, then we are adjourned.

[Whereupon, at 5:45 p.m., the subcommittee was adjourned.]

APPENDIX

July 16, 2014

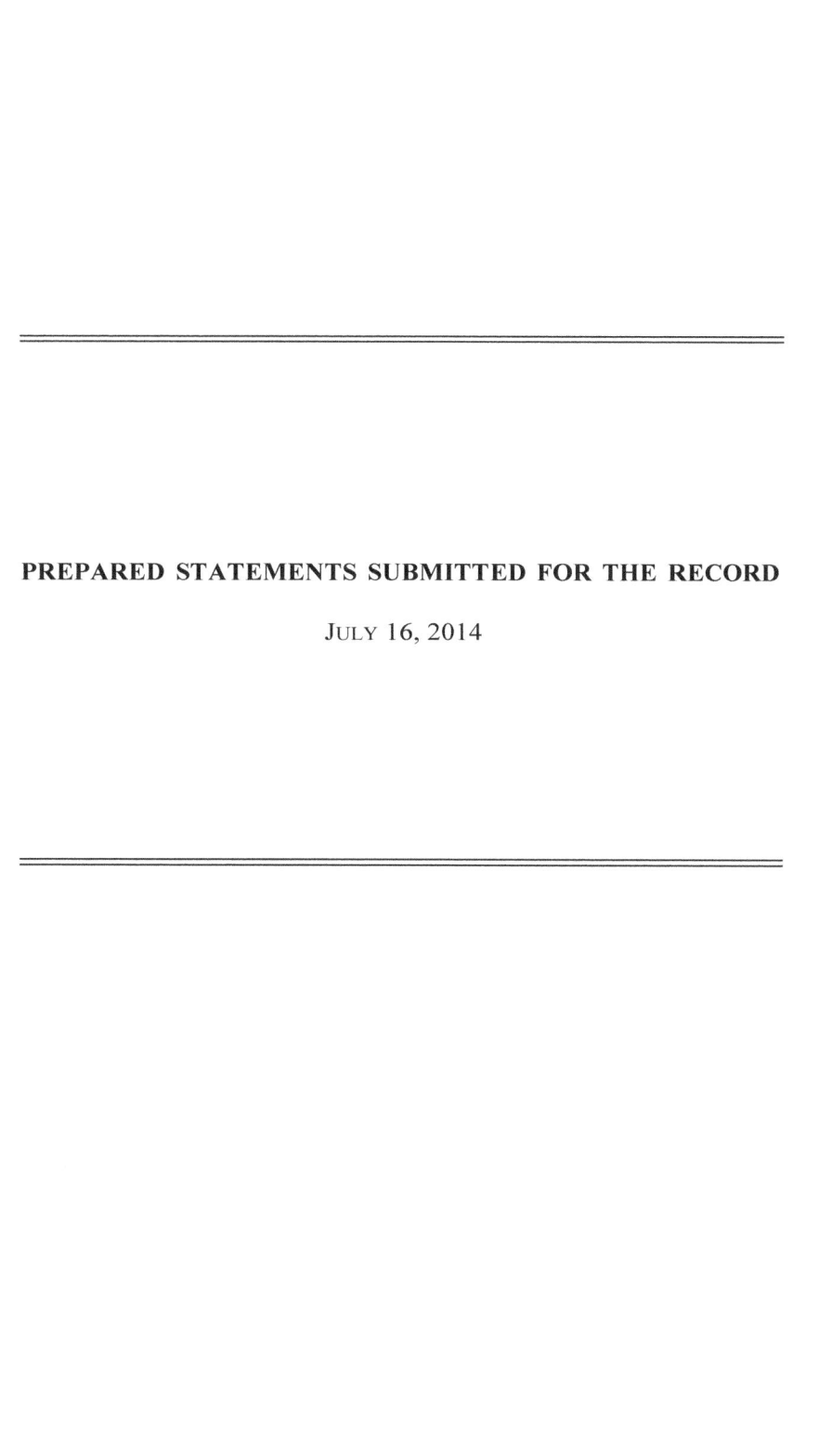

PREPARED STATEMENTS SUBMITTED FOR THE RECORD

JULY 16, 2014

Opening Remarks of the Honorable J. Randy Forbes, Chairman of the Seapower and Projection Forces Subcommittee, for the hearing on Unmanned Carrier-Launched Airborne Surveillance and Strike (UCLASS) Requirements Assessment

Today the subcommittee convenes to receive testimony on the Unmanned Carrier-Launched Airborne Surveillance and Strike (UCLASS) program.

Our first panel of distinguished guests testifying before us are:

- **Mr. Ronald O'Rourke**
  Specialist in Naval Affairs, Defense Policy and Arms Control Section
  Congressional Research Service
- **Mr. Robert Martinage**
  Former Deputy Under Secretary of the Navy
- **Mr. Shawn Brimley**
  Executive Vice President and Director of Studies
  Center for a New American Security
- **Mr. Bryan McGrath**
  Managing Director
  Ferrybridge Group LLC

Collectively, this bipartisan group has advised the U.S. Congress and Presidential campaigns, commanded a Navy large-surface combatant, drafted the 2007 Maritime

Strategy, served as Under Secretary of the Navy, served on the National Security Council Staff, and worked at various distinguished think tanks. Given their diverse background, I am confident that this bipartisan group of witnesses will be able to provide a detailed perspective to this Committee's continued work on the UCLASS program.

Our second distinguished panel, which will immediately follow this one, includes Navy and Joint Staff leaders, including:

- **Vice Admiral Paul A. Grosklags, Principal Military Deputy, Assistant Secretary of the Navy for Research, Development, and Acquisitions**
- **Mr. Mark Andress, Assistant Deputy Chief of Operations for Information Dominance**
- **Brigadier General Joseph T. Guastella, Director, Joint Requirements Oversight Council**
  Department of Defense

Gentlemen, thank you for being with us today.

We have called this hearing to discuss the Navy's UCLASS program. But before we proceed I want to be clear from the onset that I am a strong supporter of a future Carrier Air Wing that is comprised of both manned and unmanned aviation assets. The F/A-18 "Super Hornet", F-35C, EA-18G "Growler," E-2 "Hawkeye," and the UCLASS program will all be integral to ensuring our carrier fleet can continue to project power throughout the globe.

I believe the fundamental question we face is not about the utility of unmanned aviation to the future Air Wing, but the _type_ of unmanned platform that the UCLASS program will deliver and the specific _capabilities_ this vital asset will provide the Combatant Commander. Given the likely operational environment of the 2020s and beyond - including in both the Western Pacific Ocean and Persian Gulf - I believe strongly that the Nation needs to procure a UCAV platform that can operate as a long-range surveillance and strike asset in the contested and denied A2/AD environments of the future.

Unfortunately, in its current form, this Committee has concluded the UCLASS Air System Segment requirements will not address the emerging anti-access/area-denial (A2/AD) challenges to US power projection that originally motivated creation of the Navy Unmanned Combat Air System (N-UCAS) program during the 2006 Quadrennial Defense Review (QDR), and which were reaffirmed in both the 2010 QDR and 2012 Defense Strategic Guidance.

It is my determination that the disproportionate emphasis in the requirements on unrefueled endurance to enable continuous intelligence, surveillance, and reconnaissance (ISR) support to the Carrier Strike Group (CSG) would result in an aircraft design that would have serious deficiencies in both survivability and internal weapons payload capacity and flexibility. Furthermore, the cost limits for the aircraft are more consistent with a much less capable aircraft and will not enable the Navy to build a relevant vehicle that leverages readily available and mature technology. In short, developing a new carrier-based unmanned aircraft that is primarily another unmanned ISR sensor that cannot operate in medium to high-level threat environments would be a missed opportunity and inconsistent with the 2012 Defense Strategic Guidance which called for the United States to "maintain its ability to project power in areas in which our access and freedom to operate are challenged."

But the question of UCLASS is not just one of design and capability; it is also about the role and responsibility the Congress has in cultivating, supporting, and protecting military innovation. Like with the shift from cavalry to mechanized forces, sailing ships to steam-power vessels, the battleship to naval aviation, or adopting unmanned aerial vehicles in the late 1990s, ideas that initiate difficult changes and disrupt current practices are often first opposed by organizations and bureaucracies that are inclined to preserve the status quo. I believe the Congress has a unique role to help push the Department and the Services in directions that, while challenging, will ultimately benefit our national security and defense policy. I therefore intend to use this hearing today to explore not just the UCLASS program, but the broader utility a UCAV can have on the Navy's ability to continue to project power from the aircraft carrier and the implications for the power projection mission in the future if we proceed down the current course.

I again thank our two panels for being here to testify and look forward to your testimony. With that, I turn to my good friend and ranking member, the gentleman from North Carolina.

Opening Remarks for Congressman Mike McIntyre
Ranking Member, Seapower and Projection Forces Subcommittee
Unmanned Carrier Launched Airborne Surveillance and Strike
(UCLASS) System
July 16th, 2:00pm, RHOB 2212

I would like to thank the Chairman and my good friend for arranging this very important discussion today.

Thank you to all of the distinguished panelists for taking the time to come and talk with us today. We are very fortunate to have the opportunity to hear the perspectives of both an independent panel of experts as well as those of our leaders in the Department. I am confident that after today's discussion we will all have a better appreciation of the requirements and way ahead for the Unmanned Carrier Launched Airborne Surveillance and Strike (UCLASS) system. This weapon system is planned to fill a crucial capability gap within the current carrier based air wing and will play a vital role in the future of carrier based aviation. I look forward to hearing all of your opening remarks as well as the discussion to follow. Thank you again for being here and for your service to our country.

Independent Panelist:
- Mr. Robert Martinage, Senior Fellow, Center for Strategic and Budgetary Assessments
- Mr. Shawn Brimley, President and Director of Studies, Center for a New American Security
- Mr. Ron O'Rourke, Specialist in Naval Affairs, Defense Policy and Arms Control Section, Congressional Research Service

STATEMENT OF

RONALD O'ROURKE

SPECIALIST IN NAVAL AFFAIRS

CONGRESSIONAL RESEARCH SERVICE

BEFORE THE

HOUSE ARMED SERVICES COMMITTEE

SUBCOMMITTEE ON SEAPOWER AND PROJECTION FORCES

ON

UNMANNED CARRIER-LAUNCHED AIRBORNE SURVEILLANCE AND STRIKE
(UCLASS) REQUIREMENTS ASSESSMENT

JULY 16, 2014

Chairman Forbes, Ranking Member McIntyre, distinguished members of the subcommittee, thank you for the opportunity to appear before you today to discuss operational requirements for the Navy's Unmanned Carrier-Launched Airborne Surveillance and Strike (UCLASS) program.

As part of my work as a CRS naval issues analyst since 1984, I have tracked developments relating to carrier-based aviation and have written a number of reports on carrier air wing composition and programs for carrier-based aircraft.[1] In support of this testimony, I received briefings on the UCLASS program from the Navy and industry and reviewed requirements documentation for the program that the Navy provided at my request. I also reviewed the trade press reporting on the UCLASS program over the last few years.

As requested, my testimony identifies some potential key issues that the subcommittee may wish to consider in assessing operational requirements for the UCLASS program. This statement presents six such issues and makes some observations in connection with them, but it does not attempt to answer them. The six issues are by no means the only ones that might be raised, but considering them might help in forming a framework of analysis for assessing operational requirements for the UCLASS program.

A July 11, 2014, press report stated that "The Pentagon has delayed the [UCLASS] program amid ongoing reviews of the program's requirements, defense officials told Military.com."[2] My testimony discusses the UCLASS program as defined in early July 2014.

---

[1] See CRS Report 89-362 F, *Navy Carrier Aircraft in the 1990s: Critical Issues for Today*; CRS Report 90-331 F, *Navy Carrier Aircraft in the 1990s: Implications of a 12-Carrier Fleet*; CRS Report 91-88 F, *A-12 Contract Cancellation: Alternative Paths for Naval Aviation*; CRS Report 91-253 F, *Navy Carrier-Based Fighter and Attack Aircraft in the FY1992 Budget: Issues for Congress*; CRS Report 91-528 F, *Navy Carrier-Based Fighter and Attack Aircraft: House Action on the Administration's Proposed Strategy*; CRS Report 93-868 F, *Navy Carrier-Based Fighter and Attack Aircraft: Modernization Options for Congress*; CRS Report RS21488, *Navy-Marine Corps Tactical Air Integration Plan: Background and Issues for Congress* (co-authored with Christopher Bolkcom); and (during part of 2008) CRS Report RS22875, *Navy-Marine Corps Strike-Fighter Shortfall: Background and Options for Congress*. In addition, as a consequence of the sudden death in May 2009 of Christopher Bolkcom, the CRS's military aviation analyst, I restructured and maintained CRS's military aviation reports between May and November 2009—a period when Congress debated significant proposed changes to several DOD major aircraft acquisition programs. These reports covered the Air Force F-22 fighter program, the F-35 Joint Strike Fighter (JSF) program, the Navy F/A-18E/F strike-fighter and EA-18G electronic attack aircraft programs, tactical aircraft modernization (an overview report), the Air Force next-generation bomber program, the Air Force KC-X tanker aircraft program, the Air Force C-17 cargo aircraft program, the Navy V-22 Osprey tilt-rotor program, and the Navy VH-71/VXX presidential helicopter program.

[2] The article states further:

> A planned competition among defense companies has been put on hold as the Pentagon examines plans for the drone and responds to criticism from lawmakers that the initial requirements have been too narrowly configured.
>
> A formal Request For Proposal, or RFP, which had been planned for release by the Navy later this month, has been delayed by a few weeks.
>
> (Kris Osborn, "UPDATE: Pentagon Delays Navy's Carrier Drone Program," *DOD Buzz* (www.dodbuzz.com), July 11, 2014.)

Potential Shift in Strategic Eras

One issue the subcommittee may wish to consider is whether we are currently undergoing a shift in strategic eras,[3] and if so, whether and how such a shift might affect operational requirements for the UCLASS program. Actions by China starting in November 2013 that appear aimed at achieving a greater degree of control over China's near-seas region,[4] followed by Russia's seizure and annexation of Crimea in March 2014, have led to a discussion among observers about whether we are currently shifting from the familiar post-Cold War era of the last 20 to 25 years to a new and different strategic era characterized by, among other things, renewed great power competition and challenges to key aspects of the U.S.-led international order that has operated since World War II.[5] Some observers in this discussion have used the term "post-Crimea era" or "post-Crimea world."[6] I discussed aspects of this issue in my testimony for this

---

[3] The term strategic era is used here to refer to a period of time during which the structure of international relations can be said to be characterized by certain basic features. The Cold War is one example of a strategic era; the post-Cold War era (also sometimes called the unipolar moment, with the United States as the unipolar power) is another.

[4] For a summary of these actions, see CRS Report R42784, *Maritime Territorial and Exclusive Economic Zone (EEZ) Disputes Involving China: Issues for Congress*, by Ronald O'Rourke.

[5] See, for example, Anna Applebaum, "China and Russia Bring Back Cold War Tactics," *Washington Post (www.washingtonpost.com)*, December 25, 2013; Paul Miller, "China, the United States, and Great Power Diplomacy," *Foreign Policy (http://shadow.foreignpolicy.com)*, December 26 2013; Zachary Keck, "America's Relative Decline: Should We Panic? The End of the Unipolar Era Will Create New Dangers That the World Mustn't Overlook," *The Diplomat (http://thediplomat.com)*, January 24, 2013; Dan Blumenthal and Michael Mazza, "China Is Like Russia," *The Weekly Standard (www.weeklystandard.com)*, March 18, 2014; Paul D. Miller, "Crimea Proves that Great Power Rivalry Never Left Us," *New York Times (www.nytimes.com)*, March 21, 2014; March Chad Pillai, "The Return of Great Power Politics: Re-Examining the Nixon Doctrine," *War on the Rocks (http://warontherocks.com)*, March 27, 2014; Robert Killebrew, "Containing Russia and Restoring American Power," *War on the Rocks (http://warontherocks.com)*, March 27, 2014; David Roche, "West Stumbles as Autocractic Force Trumps Economics," *Reuters (www.reuters.com)*, April 1, 2014; David B. Rivkin Jr. and Lee A. Casey, "The Outlaw Vladimir Putin," *Wall Street Journal (http://online.wsj.com)*, April 8, 2014; Debidatta Aurobinda Mahapatra, "The Post-Crimea World Order," *Russia and India Report (http://in.rbth.com)*, April 14, 2014; Tom Rotnem, "10 Days That Shook the (Post-Cold War) World, *Marietta Daily Journal (http://mdjonline.com)*, April 22, 2014; Walter Russell Mead, "The Return of Geopolitics," *Foreign Affairs (www.foreignaffairs.com)*, May/June 2014; Eric A. Posner, "Sorry, America, the New World Order Is Dead," *Foreign Policy (www.foreignpolicy.com)*, May 6, 2014; Dan Blumenthal and Michael Mazza, "China and The Age of Contempt," *Foreign Policy (http://shadow.foreignpolicy.com)*, May 15, 2014; Robert Kagan, "Superpowers Don't Get to Retire," *New Republic (www.newrepublic.com)*, May 26, 2014; Walter Russell Mead, "Putin Did Americans a Favor," *Wall Street Journal (http://online.wsj.com)*, June 1, 2014; James R. Holmes, "5 Ways Europe Can Help the US Pivot," *The Diplomat (http://thediplomat.com)*, June 2, 2014; Walter Russell Mead, "For the U.S., a Disappointing World," *Wall Street Journal (http://online.wsj.com)*, June 13, 2014; James Kitfield, "The New Great Power Triangle Tilt: China, Russia Vs. U.S.," *Breaking Defense (http://breaking defense.com)*, June 19, 2014; Frank Hoffman, "No Strategic Success Without 21st Century Seapower: Forward Partnering," *War on the Rocks (http://warontherocks.com)*, July 1, 2014; David Hodges, "The Only Defense," *Commentary (www.commentarymagazine.com)*, July 1, 2014; Marc M. Wall, "The Great Eurasian Rebalancing Act," *PacNet (Pacific Forum CSIS)*, Number 52, July 7, 2014.

[6] See, for example, Jim Thomas, "How to Put Military Pressure on Russia," *Wall Street Journal (http://online.wsj.com)*, March 9, 2014; Debidatta Aurobinda Mahapatra, "The Post-Crimea World Order," Russia and India Report (http://in.rbth.com), April 14, 2014; Tom Rotnem, "10 Days That Shook the (Post-Cold War) World, *Marietta Daily Journal (http://mdjonline.com)*, April 22, 2014; "Reshaping Transatlantic Defense and Security for a Post-Crimean World," Panel remarks by NATO Deputy Secretary General Ambassador Alexander Vershbow at the Wrocław Global Forum (Poland), accessed July 2,2014, at: http://www.nato.int/cps/en/natolive/opinions_110902.htm?selectedLocale=en; Lilia Shevtsova, "Crowning a Winner

subcommittee's December 11, 2013, hearing on U.S. Asia-Pacific strategic considerations related to PLA naval forces modernization.[7]

A shift in strategic eras can lead to a reassessment of assumptions and frameworks of analysis relating to defense funding levels, strategy, missions, plans, and programs. The shift from the Cold War to the post-Cold War era led to such a reassessment in the early 1990s. This reassessment (a portion of which was carried out by the House Armed Services Committee under its chairman at the time, Representative Les Aspin) led to numerous substantial changes in U.S. defense plans and programs, including substantial changes in plans for the acquisition of carrier-based aircraft.[8] Numerous other defense programs were changed to lesser degrees or were not changed.

A shift from the post-Cold War era to a new strategic era could lead to a new reassessment of assumptions and frameworks of analysis relating to defense funding levels, strategy, missions, plans, and programs. There are some indications that elements of such a reassessment may have begun. For example, some observers, including General Philip Breedlove, the Commander of U.S. European Command, have raised the issue of whether the United States should consider halting the U.S. military drawdown in Europe, so as to respond to a more assertive Russia.[9] As another possible example, Secretary of Defense

---

in the Post-Crimea World," *American Interest (www.the-american-interest.com)*, June 16, 2014; Evan Braden Montgomery, "China's Missile Forces Are Growing: Is It Time to Modify the INF Treaty?" *The National Interest (http://nationalinterest.org)*, July 2, 2014.

[7] See Statement of Ronald O'Rourke, Specialist in Naval Affairs, Congressional Research Service, Before the House Armed Services Committee, Subcommittee on Seapower and Projection Forces, on U.S. Asia-Pacific Strategic Considerations Related to PLA Naval Forces Modernization, December 11, 2013, pp. 2 and 5-6.

[8] The shift from the Cold War to the post-Cold War era led to a shift in the Navy's formal planning emphasis away from the scenario of mid-ocean operations against Soviet naval forces during a NATO-Warsaw Pact conflict and toward operations in littoral waters against the land- and sea-based forces of countries other than Russia. This shift was formalized in a Navy/Marine Corps strategy document entitled ...*From the Sea* (the ellipse is part of the title), which was first issued in late 1992. (The text of this document is available at: http://www.au.af.mil/au/awc/awcgate/navy/fromsea/fromsea.txt.) The shift in strategic eras and in the Navy's formal planning emphasis led to numerous changes in Navy plans and programs. In terms of overall Navy force structure, the planned size of the fleet was reduced considerably. In undersea warfare, changes included the truncation of the Seawolf (SSN-21) submarine program, the initiation of the successor Virginia-class submarine program, an increased emphasis on shallow-water antisubmarine warfare (ASW) operations (including torpedoes with improved shallow-water performance), and a reduced emphasis on blue-water ASW operations. In surface warfare, the shift in planning emphasis led to the initiation of a program for a multimission destroyer (now known as the DDG-1000) with an emphasis on operations in littoral waters and land-attack operations, the initiation years later of the Littoral Combat Ship (LCS) program for addressing identified capability gaps for countering mines, small boats, and diesel-electric submarines in littoral waters. In naval aviation, changes in projected mission demands, defense spending levels and (in the case of the A-12 program) development challenges led to a broad restructuring of naval aviation acquisition programs, including the termination of the A-12 program, the halting of plans or proposals for procuring other types of carrier-based aircraft, the termination of a program to develop a new long-range air-to-air missile for carrier-based fighters, and the initiation of the F/A-18E/F program. For additional discussion of the then-emerging impact on Navy plans and programs resulting from the shift from the Cold War to the post-Cold War era, see Ronald O'Rourke, "The Future of the U.S. Navy," in Joel J. Sokolsky and Joseph T. Jockel, editors, *Fifty Years of Canada-United States Defense Cooperation, The Road From Ogdensburg*, Edwin Mellen Press, 1992 (papers delivered at "The Road from Ogdensburg: Fifty Years of Canada-U.S. Defense Cooperation," a conference held August 16-17, 1990, at St. Lawrence University, Canton, New York), pp. 289-331.

[9] Philip Ewing, "General: U.S. Should Stop European Drawdown," *Poliltico Pro Defense*, July 1, 2014. See also Steven Erlanger, "Europe Begins to Rethink Cuts to Military Spending," *New York Times (www.nytimes.com)*, March 26, 2014; Andrew Tilghman, "Spotlight Back on U.S. European Command," *Military Times*

Chuck Hagel, in his February 2014 announcement regarding the Littoral Combat Ship (LCS) program, stated in part:

> The LCS was designed to perform certain missions—such as mine sweeping and anti-submarine warfare—in a relatively permissive environment. But we need to closely examine whether the LCS has the independent protection and firepower to operate and survive against a more advanced military adversary and emerging new technologies, especially in the Asia Pacific. If we were to build out the LCS program to 52 ships, as previously planned, it would represent one- sixth of our future 300-ship Navy. Given continued fiscal restraints, we must direct shipbuilding resources toward platforms that can operate in every region and along the full spectrum of conflict.[10]

Current operational requirements for the UCLASS program reflect capability-gap analyses and an analysis of alternatives (AOA) that were done in 2009-2011 and then updated and revalidated from 2012 through April 2013.[11] This activity predates the events starting in late 2013 that have led to the discussion among observers over the possible shift in strategic eras that began in late 2013. Potential questions for Congress to consider include the following:

- Are we now undergoing a shift from the post-Cold War era to a new strategic era?

- If so, should this lead to a reassessment of assumptions and frameworks of analyses relating to defense funding levels, strategy, missions, plans, and programs?

- If so, what effect, if any, might such a reassessment have on requirements for the UCLASS program?

Cost, Schedule, and Technical Risk

A second issue the subcommittee may wish to consider is how operational requirements for the UCLASS program might affect cost, schedule, and technical risk. My comments below relate to the cost portion of this question.

Cost can include development, procurement, and life-cycle operation and support (O&S) cost, and can include consideration not only of the UCLASS program itself, but also the impact of the UCLASS program on resulting development, procurement, and life-cycle O&S costs for the remainder of the air wing. The UCLASS AOA concluded that estimated costs are to some degree sensitive to operational requirements.

(www.militarytimes.com), March 27, 2014; Peter Apps and Adrian Croft, "Crimean Pushes NATO Back to Russian Focus," *Reuters (www.reuters.com)*, March 19, 2014; Karen DeYoung, "As U.S. Ponders Next Moves on Crimea, Experts Rethink NATO's Defense Posture," *Washington Post (www.washingtonpost.com)*, March 18, 2014; Steven Erlanger, "Russia's Aggression in Crimea Brings NATO Into Renewed Focus," *New York Times (www.nytimes.com)*, March 18, 2014.

[10] Remarks by Secretary Hagel and Gen. Dempsey on the fiscal year 2015 budget preview in the Pentagon Briefing Room, February 24, 2014, accessed July 1, 2016, at: http://www.defense.gov/transcripts/transcript.aspx?transcriptid=5377.

[11] The Navy states that "Requirements for UCLASS have been fully vetted and stable since the Capabilities Development Document was approved in April 2013." (Source: Navy information paper on UCLASS program dated April 29, 2014.)

Affordability is a key performance parameter (KPP) for the UCLASS program. When I asked the Navy about the origin of the dollar value that defines the program's affordability KPP and about the origin of the FYDP funding levels for the program, the Navy replied that they were based on the results of the UCLASS AOA update and Navy discussions with industry about potential UCLASS costs, plus lessons learned from the Navy's X-47B UCAS-D (Unmanned Combat Air System—Demonstration) effort. The AOA update reflects the program's current operational requirements, and the Navy stated that its discussions with industry about potential UCLASS program costs also reflected the program's current operational requirements. The Navy also stated that it could not locate a document showing how the dollar value for the affordability KPP was established, and that the figure appears to reflect a judgment made by senior Navy officials.

Since the AOA update and the Navy's discussions with industry both reflect the program's current operational requirements, basing the UCLASS program's affordability KPP to a large degree on these two things can help ensure that the affordability KPP and the FYDP funding levels are realistic for the program as currently defined, which in turn can reduce the likelihood of cost growth in the program as currently defined. At the same time, in the context of a debate over operational requirements for the UCLASS program, this approach to establishing the program's affordability KPP can produce a definition of affordability that can be viewed as circular to some degree, because it can be understood as saying, in essence, "What is affordable is a program with the current operational requirements." Since estimated costs for the UCLASS program are to some degree sensitive to operational requirements, a definition of affordability that is to some degree circular in nature in relation to operational requirements has the potential for being invoked as a rhetorical device for discouraging or closing down debate on operational requirements.

Other Navy acquisition programs in the past have used different approaches for defining their affordability cost targets. For example, the original unit procurement cost target for the DDG-51 destroyer program was calculated so that procuring DDG-51s at projected rates would not require more than a certain percentage of projected future shipbuilding budgets.[12] The affordability of the Seawolf (SSN-21) attack submarine program was similarly discussed by the Navy in terms of how procuring SSN-21s at projected rates would not require more than a certain percentage of projected future shipbuilding budgets.[13] A similar approach for the UCLASS program might define affordability based on a percentage of projected funding for carrier-based aircraft acquisition.

Outcomes in Potential Operational Scenarios

A third issue the subcommittee may wish to consider is how operational requirements for the UCLASS program might affect estimated outcomes in future operational scenarios. These outcomes can be estimated through computer modeling and wargames. Key metrics to examine can include things like

---

[12]See Jan Paul Hope and Vernon E. Stortz, "Warships and Cost Constraints," *Naval Engineers Journal*, March 1986: 41-52, particularly the section entitled "Setting the Cost Constraints for DDG-51" on page 43.

[13] See CRS Report 93-10 F, *Navy Centurion Attack Submarine: What is Affordable?* by Ronald O'Rourke; and CRS Report 94-643 F, *Navy New Attack Submarine (NSSN) Program: Is It Affordable?* by Ronald O'Rourke. (Both reports are out of print and available from the author. Centurion and New Attack Submarine [NSSN] were early names for what is now known as the Virginia-class submarine.) See also Statement of Ronald O'Rourke, Specialist in National Defense, Congressional Research Service, Before the House National Security Committee, Subcommittee on Military Procurement, Hearing on Submarine Acquisition Issues, March 16, 1995, pp. 20-22. The Seawolf program's affordability in terms of percentage of projected future shipbuilding budgets was mentioned by the Navy in testimony to Congress several times in the mid-1980s.

probability of victory or mission fulfillment, time needed to achieve victory or mission fulfillment, and U.S. and coalition losses incurred on the way to victory or mission fulfillment.

The specific tactical situations that were examined in the UCLASS AOA are related to the program's current operational requirements. Assessing alternative operational requirements for the UCLASS program could involve examining potential outcomes in other tactical situations that may not have been considered in the AOA. A broader analysis might examine how changes in UCLASS operational requirements might affect estimated outcomes in campaign-level, force-on-force situations, rather than in specific tactical situations.

Threat Assessments

A fourth issue the subcommittee may wish to consider is how operational requirements for the UCLASS program relate assessments of potential future adversary capabilities, including capabilities for anti-surface warfare, air warfare (e.g., fifth-generation aircraft), and air defense. Specific questions that Congress may want to consider include the following:

- How does the growth potential of the baseline UCLASS aircraft design relate to projected improvements over time in adversary capabilities?

- How sensitive are operational requirements for the UCLASS program to changes in assessments of potential future adversary capabilities? How much uncertainty or potential for change is there in these threat assessments?

Technology Paths

A fifth issue the subcommittee may wish to consider is how operational requirements for the UCLASS program might affect potential technology paths for future systems and capabilities. The Navy's current vision statement for naval aviation, entitled *Naval Aviation Vision 2014-2025*, incorporates the UCLASS program as it is currently defined.[14] Specific questions that Congress may consider include the following:

- How might the Navy's current long-range vision for naval aviation be affected, if at all, by the potential shift in strategic eras discussed above?

- How might operational requirements for the UCLASS program affect the baseline UCLASS design or the technologies developed for the UCLASS program, and what effect might this in turn have on opening up, preserving, or encumbering potential pathways for achieving the Navy's current long-term vision for naval aviation or potential alternatives to that current long-term vision?

Other Countries

A sixth issue the subcommittee may wish to consider is how operational requirements for the UCLASS program might affect the behavior of other countries. Specific questions that Congress may consider include the following:

---

[14] *Naval Aviation Vision 2014-2025*, pp. 9, 23, 36, 62 (graphic), 63, 70.

- What impact might operational requirements for the UCLASS program have in terms of imposing costs on potential adversaries, or on persuading potential adversaries from taking certain courses of action?[15]

- What impact might operational requirements for the UCLASS program have in terms of reassuring U.S. allies and partners regarding U.S. intentions and resolve?

In the UCLASS briefing materials provided to me by the Navy, I did not notice an analysis of how operational requirements for the UCLASS program might affect behavior of other countries. It is possible that such an analysis is presented in other UCLASS program documents.

---

[15] Regarding cost-imposing strategies, a 2012 Marine Corps report on potential future U.S. amphibious warfare capabilities, for example, states:

> The assurance of sustained littoral access presents a cost-imposing deterrent to would-be opponents, and is a hedge against unforeseen requirements in a rapidly changing security environment.... For conventional deterrence, the ability to conduct forcible entry is a cost-imposing strategy that serves to deter regional powers from provocative behavior..... Littoral maneuver enables modern amphibious doctrine by avoiding attacking frontally onto a defended beach. It presents a cost-imposing asymmetry for the enemy that forces him to defend many places at once.
>
> (*Naval Amphibious Capability in the 21st Century, Strategic Opportunity and a Vision for Change, Report of the Amphibious Capabilities Working Group*, April 27, 2012, pp. 12 and S-9.)

The origin of this report is explained as follows:

> The January 2012 publication of *Sustaining U.S. Leadership: Priorities for 21st Century Defense* is premised on the interests of a maritime nation with global responsibilities and an imperative to lead. Simultaneously, it marks a strategic inflection point after a decade of sustained operations ashore. The U.S. will continue to lead, but will do so with new capabilities, in new places, with an eye toward new threats. Recognizing this changing tide, and the opportunities contained within it, Marine Corps senior leadership convened an Amphibious Capabilities Working Group (ACWG) to step back from the momentum created by current operations and programs. Service leadership, through the forum of the Marine Requirements Oversight Council (MROC), demanded a comprehensive review of Service concepts and capabilities through the lens of national strategic priorities and the newly emerging security environment." (Page 7)

As another example, a 2013 Carnegie Endowment report on China's military and the U.S.-Japan alliance in 2030 states that

> the compounding nature of small changes is only one reason why the long-term perspective of net assessments is useful. There are several other reasons why policymakers need to consider the long term. First, of course, it puts various aspects of a particular conflict in perspective. Although the term "Cold War" certainly described one dimension of the interaction between the United States and the Soviet Union, understanding it as a "long-term competition" provided an entirely different—and sometimes very useful—strategic perspective. Some argue, for example, that while bombers were not the best and most efficient nuclear delivery mechanism during the Cold War, they played a little understood role in U.S. strategy: "By continually adding new planes and cruise missiles to the U.S. arsenal over the past three decades, Washington has forced Moscow to invest heavily in such purely defensive weapons as antiaircraft missiles. Over the years, this investment has been more expensive for the Soviet Union, and at the same time, it is less threatening to the United States than Soviet investment in tanks, ballistic missiles, or other offensive weapons." In fact, when thinking about conflicts as a form of long-term competition, a variety of potential cost-imposing or competitive strategies becomes potentially useful. Thus, in the long-term competition in the Western Pacific, though China's rate of military modernization might be a reason for concern, it is certainly not a reason for panic; there are both military and nonmilitary actions the United States, Japan, and others can undertake and strategies they can adopt that can improve their relative positions in the long term.
>
> (Michael D. Swaine, *et al, China's Military & the U.S.-Japan Alliance in 2030, A Strategic Net Assessment*, Carnegie Endowment for International Peace, 2013, pp. 11-12.)

Mr. Chairman, this concludes my statement. Thank you again for the opportunity to testify, and I will be pleased to respond to any questions the subcommittee may have.

Ronald O'Rourke

Specialist in Naval Affairs, Defense Policy and Arms Control Section

Congressional Research Service

Mr. O'Rourke is a Phi Beta Kappa graduate of the Johns Hopkins University, from which he received his B.A. in international studies, and a valedictorian graduate of the University's Paul Nitze School of Advanced International Studies, where he received his M.A. in the same field.

Since 1984, Mr. O'Rourke has worked as a naval analyst for the Congressional Research Service of the Library of Congress. He has written numerous reports for Congress on various issues relating to the Navy. He regularly briefs Members of Congress and Congressional staffers, and has testified before Congressional committees on several occasions.

In 1996, Mr. O'Rourke received a Distinguished Service Award from the Library of Congress for his service to Congress on naval issues.

Mr. O'Rourke is the author of several journal articles on naval issues, and is a past winner of the U.S. Naval Institute's Arleigh Burke essay contest. He has given presentations on Navy-related issues to a variety of audiences in government, industry, and academia.

July 16, 2014

STATEMENT BEFORE THE HOUSE ARMED SERVICES SUBCOMMITTEE ON SEAPOWER AND PROJECTION FORCES ON THE UNMANNED CARRIER-LAUNCHED AIRBORNE SURVEILLANCE AND STRIKE (UCLASS) REQUIREMENTS ASSESSMENT

By Robert Martinage
Senior Fellow
Center for Strategic and Budgetary Assessments

Chairman Forbes, Ranking Member McIntyre, and members of this distinguished committee, thank you for the opportunity to share my views on UCLASS requirements. I'd like to commend the committee for taking an active interest in what is one of the most important force development issues facing DoD in general and the US Navy in particular. I've studied this issue for many years and from several different vantage points—first as an analyst at the Center for Strategic and Budgetary Assessments (CSBA), then as a senior civilian in both the Office of the Secretary of Defense (OSD) and the Navy Secretariat—and I consider it to be a harbinger of DoD's ability to transform how it projects power to meet emerging challenges.

Ironically, the ongoing debate over carrier-based unmanned air system (UAS) roles and missions is analogous to the debate during the interwar years over the role of the nascent aircraft carrier. At the time, the dominant view within the Navy was that carriers should provide airborne surveillance for battleships rather than serve as an independent striking arm of the fleet. The Chief of Naval Operations at the time, Admiral Benson, made the remark, "the Navy doesn't need airplanes. Aviation is just a lot of noise."[1] Reflecting this deeply ingrained cultural view, the program for the November 29, 1941 Army-Navy football game prominently featured a classic bow-on picture of the *USS Arizona* plunging through a huge ocean swell with the caption: "Despite the claims of air enthusiasts, no battleship has yet been sunk by bombs." Just one week later, Japanese carrier-based aircraft sank the *Arizona* pier-side in Pearl Harbor. In the years that followed, American aircraft carriers rapidly became the linchpin of the war in the Pacific.

Furthermore, the aircraft carrier has remained a crucial means of U.S. global power projection ever since, providing a mobile sea-base that can be positioned wherever needed. It has maintained its strategic effectiveness over the past 70 years because of the adaptability afforded by its embarked air wing—from torpedo- and dive-bombers at the Battle of the Coral Sea to F/A-18 strike fighters in Operation Enduring Freedom.

---

[1] William F. Trimble, *Admiral William A. Moffett: Architect of Naval Aviation* (Washington, DC: Smithsonian Institution Press, 1994), p. 71.

UCLASS should be the logical next step in the evolution of the carrier air wing. Near-term decisions on UCLASS' system performance requirements, however, will have a profound impact on its future operational utility. Poor decisions could eventually be reversed at higher cost—in terms of time, operational risk, and resources. However, given current budget constraints, it is likely that the nation would be saddled with these consequences for years to come. It is imperative to get the requirements right the first time, and this is accomplished in part by focusing on meeting emerging power projection challenges that the Intelligence Community anticipates will intensify and proliferate over the coming decades—not solely on meeting current operational demands.

An assessment of UCLASS requirements should begin with a very simple question: what core operational problem should the UCLASS be designed to solve? The dominant answer within the Navy, reflected in the UCLASS draft request for proposal (RFP), are the needs to maintain continuous maritime domain awareness (MDA) around the Carrier Strike Group (CSG) as well as identify targets for attack by relatively short-range, manned fighters. An alternative answer, one I espouse firmly, is that the more pressing problem is maintaining the Navy's ability to project power from the sea when: 1) carriers are compelled to standoff at considerable distance (e.g., in excess of 1,000 miles) from an adversary's territory due to emerging anti-access/area-denial (A2/AD) threats such as long-range anti-ship cruise missiles (ASCMs) and anti-ship ballistic missiles (ASBMs); and 2) it is necessary to find and destroy fixed and mobile/relocatable targets defended by modern integrated air defenses (IADS).

The Current Draft RFP: UCLASS as a "Spotter" for Manned Aircraft in Low-Mid Threat Environments

Driven by the perceived need to sustain continuous MDA around the CSG, including overnight while the deck is "closed," the draft RFP contains a derived threshold requirement for an unrefueled endurance of about 14 hours. The latter is required to sustain two continuous "24–7" intelligence, surveillance, and reconnaissance (ISR) orbits at a required radius of 600 nm from the carrier without violating the carrier's 12-hour "deck day" or requiring aerial tanking support.

The opportunity cost of 14 hours of unrefueled endurance, however, comes in the form of *permanent* design trades that significantly reduce the aircraft's survivability and payload carriage/flexibility—attributes needed to perform ISR *and* precision strike roles in A2/AD environments. These foregone capabilities cannot be "bought back" later or added to future UCLASS variants. Claims that "threshold growth" and "objective" requirements in the draft RFP will place competitive pressure on industry to enhance survivability and payload attributes are largely a chimera. As a matter of physics, absent a break-through in engine technology, it is impossible to achieve 14 hours of unrefueled endurance with an aircraft sized to operate from an aircraft carrier without making choices about its shape and propulsion path that constrain passive radar signature reduction (i.e., stealth) and potential and internal weapon carriage capacity (including both the numbers and types of weapons carried). While it is true that a few hours of endurance could be gained by installing internal fuel tanks in the UCLASS' bomb bay, it does not significantly expand the design trade space. Similarly, while additional payload could be carried externally with a significant reduction in endurance, it would also make the aircraft even less survivable in contested air space. Simply put, meeting the threshold requirement of 14 hours of unrefueled endurance necessarily results in sacrificing survivability, weapons carriage/flexibility, and growth margins for future mission payloads (i.e., space, weight, power, and cooling allowances) and there is no technologically viable "growth path" for restoring them.

Perhaps this opportunity cost would be acceptable if there were a compelling operational justification for ~14 hours of unrefueled endurance—but there is not. It is worth noting that an aircraft with 8–10 hours of unrefueled endurance, flying at high subsonic speeds, would have roughly *three times* the combat radius of the F/A-18E/F or F-35C. To put this in operational perspective, that same 8–10 hour endurance aircraft could launch from a carrier positioned 1,000 miles away from an area of interest (the range of the Chinese DF-21D ASBM), loiter on-station for 3–4 hours, then recover onboard the carrier with reserve fuel as a safety margin.

When factoring in aerial refueling—an Air Force-supplied resource typically available to carrier-based aircraft in wartime—the 14-hour unrefueled endurance threshold requirement makes even less sense. The same 8–10 hour endurance aircraft could take off from a carrier positioned virtually any distance from a prospective adversary, refuel in transit and on ingress to the combat zone at a safe stand-off range for the tanker, remain on-station for 5–7 hours, cycle to the tanker and back to operational station *multiple times*, and eventually recover to either a carrier already in the region or its original home carrier. The marriage of unmanned operations and aerial refueling would enable the aircraft carrier to launch missions from intercontinental range in response to surprise aggression as well as to sustain persistent surveillance and strike operations from "access-insensitive" distances. For these reasons, automated aerial refueling (AAR) should be a threshold requirement for any carrier-based UAS program.

For survivability, it is important that UCLASS' level of radar cross section (RCS) reduction anticipates that future fire control radars will provide higher targeting resolution at lower frequencies by harnessing more powerful data processing techniques. It is also critical to address today's lower frequency acquisition and early warning radars, which have proliferated widely and are already integrated into the air defense networks of several prospective adversaries. Tracks generated by early warning radars will not only enable more efficient cued searches by fire control radars, but they could also be used to vector air defense fighters to intercept friendly aircraft. The technology required to achieve the level of RCS reduction required across the full threat radar frequency spectrum associated with 2025+ air defenses is both mature and affordable. Despite arguments to the contrary, "stealth" is not a primary driver of aircraft cost. While there are marginal costs associated with radar-absorbent edges and coatings, as well as sensor aperture integration, stealth is fundamentally a choice about the air vehicle's shape and propulsion path.

I am not aware of any mission- or campaign-level analysis showing that a threshold payload requirement of 1,000 lbs. is sufficient for a carrier-based UAS. Given the number of weapons required to both saturate an adversary's short-range air defenses and hit multiple aim points, 1,000 lbs. of payload (e.g., four small diameter bombs) is clearly inadequate to defeat most relevant targets such as coastal defense cruise missile sites, air defense radars, missile launchers, or enemy surface ships. In addition, the Navy has given scant consideration to the *types* of weapons that UCLASS should be able to accommodate. Since even a sufficiently stealthy UCLASS would be vulnerable if it approached too close to heavily defended targets, it should be able to carry stand-off weapons such as the Joint Standoff Weapon (JSOW), Long-Range Anti-Ship Missiles (LRASMs), and/or Joint Strike Missiles (JSMs). As adversary air defense radars become more capable over time, as they inevitably will, UCLASS could maintain its overall survivability by employing stand-in electronic attack techniques, as well as by finding and engaging targets at greater stand-off distances. The latter, however, will require more capable sensors and longer range weapons, and that kind of adaptability must be designed in upfront with margins for space, weight, power, and cooling.

Finally, a carrier-based UAS optimized for ISR missions in relatively benign threat environments would be a redundant capability. The Navy is already procuring more than 60 MQ-4C *Tritons* designed specifically to provide broad-area maritime surveillance for deployed CSGs. The MQ-4C, augmented by the MQ-8B/C *Firescout*, which can operate from any air-capable ship in the fleet, could provide MDA around the CSG more effectively and affordably. For persistent ISR coverage over land in low-to-medium threat environments, the joint force has more than enough capacity with the currently projected fleet of RQ-4 *Global Hawks*, MQ-1C *Gray Eagles*, and MQ-9 *Reapers*. With 30–40 hours of unrefueled endurance, RQ-4s and extended-range MQ-9s could access any area of interest with a very high degree of basing flexibility.

A Balanced Design: Carrier-Based UAS as an Independent Striking Arm and Enabler of Manned Fighter Squadrons

A balanced carrier-based UAS design would, in this order of priority:
> 1) Achieve the minimum level of broadband, passive signature reduction required to locate priority targets with onboard sensors and engage them effectively with available weapons without being destroyed by modern air defenses;
> 2) Provide sufficient unrefueled endurance to reach target areas when the carrier is standing off at least 1,000 miles with allowances for indirect routing, maneuvering and loiter time; and
> 3) Once the above two conditions are satisfied, carry as much payload and as many types of weapons as possible while still conforming to carrier-deck size constraints.

A more balanced carrier-based UAS could have, for example, an unrefueled endurance of 8–10 hours (which translates to a combat radius of ~1,700–2,000 nm from the carrier or tanker); 24–48 hours of mission endurance with air-to-air refueling; broadband/all-aspect RCS reduction sufficient to find and engage defended targets; and the ability to carry 3,000–4,000 lbs. of strike payload internally (roughly what the F-35C can carry), including a variety of direct and stand-off weapons (see Figure 1).

Figure 1—Comparison of Draft RFP and "Balanced" UCLASS Designs

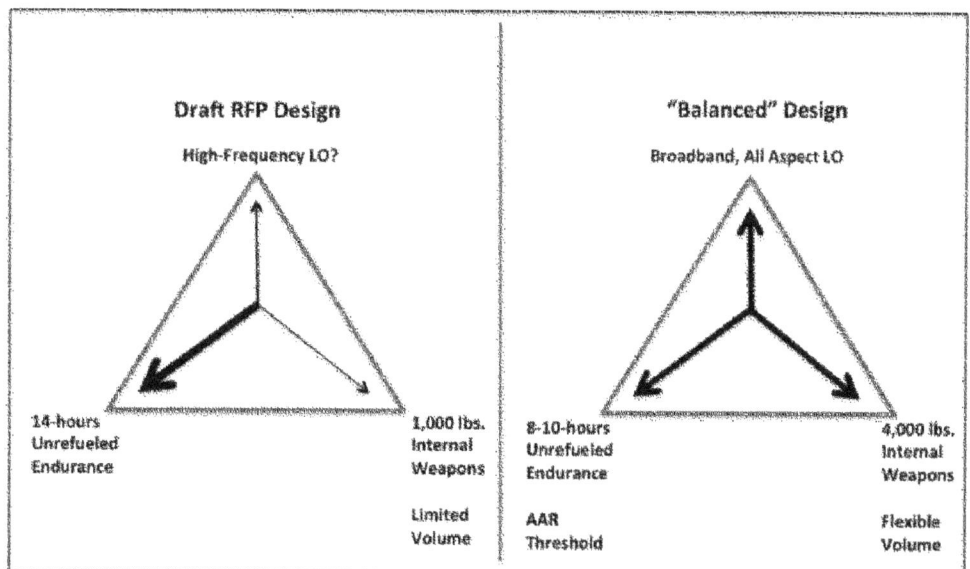

A "balanced" UCLASS could serve as an independent, long-range surveillance and striking arm of the aircraft carrier in A2/AD environments anticipated for 2025 and beyond. With aerial tanking support, it could respond globally to short notice aggression regardless of the carrier's initial location. Once the carrier was in position, outside of the densest A2/AD threats, it could contribute to a sustained precision strike campaign against an adversary's fixed and mobile targets. As part of the joint force, it could focus on coastal/shallow inland targets and naval targets such as surface action groups (SAGs). Taking advantage of its ultra-long mission endurance, it could be especially effective in hunting down and destroying mobile or relocatable targets over wide geographic areas.

A balanced UCLASS could also serve as a powerful enabler of manned carrier-based aircraft, in which the nation has invested billions of dollars, and do so in ways other than just finding targets in relatively permissive environments. With onboard fuel storage of about 20,000 lbs., it would be a very efficient aerial refueler for relatively short-range manned fighters. With its very low RCS, it could employ low-power, stand-in jamming techniques to improve the survivability of the F-35C—and to a lesser degree the F/A-18E/F—in higher-end threat environments.

Key Changes Required to the Draft UCLASS RFP

The opportunity cost of 4–6 hours of additional unrefueled endurance (14 vice 8–10 hours) as reportedly set forth in the draft UCLASS RFP would result in a dramatic reduction in strike capacity, a significant increase in air vehicle vulnerability, and reduced growth potential (i.e., lower margins for space, weight, power, and cooling).

To "fix" the draft RFP, five critical changes are needed to *threshold* requirements:
- Reduce unrefueled endurance from ~14 hours to 8–10 hours;
- Add automated aerial refueling (give and receive) and 24–48 hours refueled mission endurance as threshold requirements;
- Increase internal weapon payload from 1,000 lbs. to 3,000–4,000 lbs.;
- Establish weapon bay volume requirements to carry specified current and future standoff weapons (e.g., JSOW and JSM); and
- Require all-aspect, broadband RCS reduction at levels sufficient to address 2025–2035 air defense threats.

Looking Ahead to the Carrier Air Wing of 2025 and Beyond

The Navy intends to initiate development of another manned, supersonic fighter, the F/A-XX, to begin replacing older F/A-18s as they reach their end of service life in the late 2020s. The initial request for information from industry, which was clearly skewed toward a manned replacement, was released in 2012, and preparations are underway to initiate an Analysis of Alternatives (AoA). Putting aside the financial and political feasibility of concurrent fighter programs (F-35C and F/A-XX), especially given the cost and technical challenges still facing the F-35C, it is not at all clear that the future carrier air wing should be dominated by a mix of manned fighters with very limited mission endurance and combat radius.

The Navy's F/A-18 replacement plan and the draft UCLASS RFP both reflect a mindset that values unmanned aircraft as an appendage to the carrier air wing—not an integral part of it. Rather than thinking about 4–6 UAS per carrier, serious consideration should be given to fielding 1–2 squadrons per operational carrier in the 2020s, which would mean displacing manned aircraft, and thus, would prompt cultural and bureaucratic resistance within the naval aviation community. This would not only allow the carrier to serve as a flexible, global surveillance-strike platform, it would also result in significant lifecycle-cost avoidance. The Navy currently buys roughly enough of a specific type-model series of aircraft to outfit all 10 air wings so pilots can train year-round, whether they are deployed or stationed ashore. With UAS, there is no need to train pilots, so the Navy would only need to buy the number required to equip the maximum number of deployable carriers and generally fly those aircraft only when deployed. As a result, compared to manned aircraft, the Navy could buy about half as many carrier-based UAS and fly them less than half as often, generating significant savings in procurement, as well as operations and maintenance. As called for in the House version of the fiscal year 2015 National Defense Authorization Act, quantitative analysis at the campaign-level is needed across a wide-range of representative scenarios set in the 2025–2035 timeframe to determine the best composition of the future carrier air wing. Given its potential advantages in survivability, mission endurance, and life-cycle costs, a balanced UCLASS would likely perform very well.

Conclusion

There is no question that the nation needs a carrier-based unmanned aircraft. The relevant question is: *what kind* of aircraft? A system optimized for sustaining persistent ISR coverage in relatively benign threat environments is redundant and does not address the core operational problems facing naval aviation: the intensifying "anti-navy" threats that will push the carrier farther away from target areas and networked air defenses that will make non-stealthy aircraft increasingly vulnerable to detection and attack. Unless these threats are addressed, carrier aviation, which has been the heart-and-soul of America's maritime power-projection capability since World War II, may be progressively relegated to the sidelines in future conflicts.

To preserve the aircraft carrier's strategic relevance over the next several decades, the Navy needs to develop and field a carrier-based UAS with:

- Ultra-long refueled mission endurance to respond rapidly to future contingencies and sustain persistent surveillance-strike operations from carriers positioned outside of A2/AD threat range;
- Survivability sufficient to find and engage, using onboard sensors and weapons, fixed and mobile/relocatable targets defended by modern air defenses;
- Unrefueled combat radius sufficient to range the depth and breadth of the battlespace from tankers standing off outside of enemy surface-to-air missile and fighter coverage; and
- As much payload carriage and flexibility as possible to neutralize adversary targets rapidly, minimize the need to return to the carrier to rearm, and hold at risk as many classes of targets as possible.

Unfortunately for the Navy and the nation, that air vehicle is not the one currently called for in the draft UCLASS RFP.

About the Center for Strategic and Budgetary Assessments

The Center for Strategic and Budgetary Assessments (CSBA) is an independent, nonpartisan policy research institute established to promote innovative thinking and debate about national security strategy and investment options. CSBA's goal is to enable policymakers to make informed decisions on matters of strategy, security policy and resource allocation. CSBA provides timely, impartial and insightful analyses to senior decision makers in the executive and legislative branches, as well as to the media and the broader national security community. CSBA encourages thoughtful participation in the development of national security strategy and policy, and in the allocation of scarce human and capital resources. CSBA's analysis and outreach focus on key questions related to existing and emerging threats to US national security. Meeting these challenges will require transforming the national security establishment, and we are devoted to helping achieve this end.

Robert Martinage

Senior Fellow, Center for Strategic and Budgetary Assessments

Areas of Expertise

Strategy, Force Planning, Long-Term Military Competition, Future Warfare, Special Operations, Air-Space, Naval Warfare

Biography

Mr. Martinage recently returned to CSBA after five years of public service in the Department of Defense (DoD). While performing the duties of the Under Secretary of Navy, he led development of the Department of the Navy's FY 2014/2015 budgets and represented the Department during the Strategic Choices and Management Review, as well as within the Defense Management Action Group (DMAG). From 2010-2013, Mr. Martinage served as the Deputy Under Secretary of the Navy, providing senior-level advice on foreign and defense policy, naval capability and readiness, security policy, intelligence oversight, and special programs. Appointed Principal Deputy Assistant Secretary of Defense for Special Operations, Low-Intensity Conflict and Interdependent Capabilities in the Office of the Secretary of Defense (OSD) in 2009, Mr. Martinage focused on special operations, irregular warfare, counter-terrorism, and security force assistance policy. He also led a two-year, DoD-wide effort to develop an investment path for a future long-range strike "family of systems."

Prior to his government service, Mr. Martinage was employed at CSBA where he was responsible for carrying out a broad research program on defense strategy and planning, military modernization, and future warfare for government, foundation, and corporate clients. He has more than 14 years of experience designing, conducting, and analyzing over 40 wargames and numerous studies focused on conventional operations in high-end threat environments, special operations, irregular warfare, and strategic deterrence and warfare.

Mr. Martinage holds a MA from The Fletcher School of Law & Diplomacy with concentrations in International Security Studies, Southwest Asia, International Negotiation & Conflict Resolution, and Civilization & Foreign Affairs. Mr. Martinage earned his BA cum laude from Dartmouth College in Government with concentrations in International Relations and Political Theory & Public Law.

69

DISCLOSURE FORM FOR WITNESSES
CONCERNING FEDERAL CONTRACT AND GRANT INFORMATION

INSTRUCTION TO WITNESSES: Rule 11, clause 2(g)(5), of the Rules of the U.S. House of Representatives for the 113th Congress requires nongovernmental witnesses appearing before House committees to include in their written statements a curriculum vitae and a disclosure of the amount and source of any federal contracts or grants (including subcontracts and subgrants) received during the current and two previous fiscal years either by the witness or by an entity represented by the witness. This form is intended to assist witnesses appearing before the House Committee on Armed Services in complying with the House rule. Please note that a copy of these statements, with appropriate redactions to protect the witness's personal privacy (including home address and phone number) will be made publicly available in electronic form not later than one day after the witness's appearance before the committee.

Witness name: Robert Martinage

Capacity in which appearing: (check one)

_X_ Individual

___ Representative

If appearing in a representative capacity, name of the company, association or other entity being represented:

FISCAL YEAR 2014

federal grant(s)/ contracts	federal agency	dollar value	subject(s) of contract or grant
WHS	DOD/ONA	$1,364,000	Assessments/analysis, wargames, and briefings on international security environment, strategic challenges, future warfare, and portfolio rebalancing
DLA Acquisition Directorate	National Defense University	$87,000	Secretary of Defense Corporate Fellows Program Orientation
MOBIS	Department of the Navy	$121,000	Portfolio rebalancing

FISCAL YEAR 2013

federal grant(s)/ contracts	federal agency	dollar value	subject(s) of contract or grant
WHS	DOD/ONA	$1,200,000	Assessments/analysis, wargames, and briefings on international security environment, strategic

			challenges, future warfare, and portfolio rebalancing
DLA Acquisition Directorate	National Defense University	$84,000	Secretary of Defense Corporate Fellows Program Orientation
MOBIS	Army War College	$121,000	Portfolio rebalancing
MOBIS	National Commission on the Structure of the Air Force	$75,000	Portfolio rebalancing
DARPA	DARPA	$175,000	Research and analysis

FISCAL YEAR 2012

Federal grant(s) / contracts	federal agency	dollar value	subject(s) of contract or grant
WHS	DOD/ONA	$1,800,000	Assessments/analysis, wargames, and briefings on international security environment, strategic challenges, future warfare, and portfolio rebalancing
DLA Acquisition Directorate	National Defense University	$80,000	Secretary of Defense Corporate Fellows Program Orientation
CTTSO SETA	OASD (SO/LIC)	551,000	Future requirements and visioning

Federal Contract Information: If you or the entity you represent before the Committee on Armed Services has contracts (including subcontracts) with the federal government, please provide the following information:

Number of contracts (including subcontracts) with the federal government:

Current fiscal year (2014): _____3_____ ;
Fiscal year 2013: _____5_____ ;
Fiscal year 2012: _____3_____ .

Federal agencies with which federal contracts are held:

Current fiscal year (2014): _____3_____ ;
Fiscal year 2013: _____4_____ ;
Fiscal year 2012: _____2_____ .

List of subjects of federal contract(s) (for example, ship construction, aircraft parts manufacturing, software design, force structure consultant, architecture & engineering services, etc.):

Current fiscal year (2014):___Research and analysis_____;
Fiscal year 2013:_____Research and analysis_____;
Fiscal year 2012:_____Research and analysis_____.

Aggregate dollar value of federal contracts held:

Current fiscal year (2014): ___$1,572,000_____;
Fiscal year 2013:_____$1,655,000_____;
Fiscal year 2012: _____$2,431,000_____.

Federal Grant Information: If you or the entity you represent before the Committee on Armed Services has grants (including subgrants) with the federal government, please provide the following information:

Number of grants (including subgrants) with the federal government:

      Current fiscal year (2014):_____;
      Fiscal year 2013:_____;
      Fiscal year 2012:_____.

Federal agencies with which federal grants are held:

      Current fiscal year (2014):_____;
      Fiscal year 2013:_____;
      Fiscal year 2012:_____.

List of subjects of federal grants(s) (for example, materials research, sociological study, software design, etc.):

      Current fiscal year (2014):_____;
      Fiscal year 2013:_____;
      Fiscal year 2012:_____.

Aggregate dollar value of federal grants held:

      Current fiscal year (2014):_____;
      Fiscal year 2013:_____;
      Fiscal year 2012:_____.

CONGRESSIONAL TESTIMONY

Aim Higher: Elevating the Debate Over Unmanned Carrier Aviation
Prepared Statement of Shawn Brimley

Center for a New American Security

July 16, 2014
Testimony before the House Armed Services Committee
Subcommittee on Seapower and Projection Forces
Hearing on the Unmanned Carrier-Launched Airborne Surveillance and Strike (UCLASS) Requirements Assessment

Prepared Statement of Shawn Brimley
Executive Vice President and Director of Studies, Center for a New American Security

Thank you Chairman Forbes and Ranking Member McIntyre for the opportunity to testify and submit this written statement for the record.

The issue of when and how the U.S. Armed Forces fully integrates unmanned and increasingly autonomous global surveillance and strike platforms into its inventory is one of the most important issues facing the Department of Defense. I am concerned that DOD is not aiming high enough to ensure the United States retains its hard won military-technical dominance in the very challenging period ahead.

We are in the opening phases of a discontinuous shift in military affairs, one in which the transition to a world where many nations will have access to unmanned and increasingly autonomous systems will cause major disruptions in the military balance of power.[1] This emerging war fighting regime will evolve during an era in which the United States will face intense, asymmetric military-technical competition from rising powers and even non state actors aiming to exploit the Achilles Heel of U.S. defense strategy—our utter dependence on extended mobilization times and permissive operating environments. As a strategic matter, it is critical that the Department of Defense lead, not follow, in terms of technical, conceptual, and operational innovation. Budgets are tight and resources scarce, but we must think big, act boldly, and aim for nothing less than technological dominance in the decade ahead. To do anything less is to court the erosion of U.S. military power and a decline in our ability to shape the future.

Therefore, the question of how the U.S. Navy approaches the unmanned carrier-launched airborne surveillance and strike (UCLASS) program is not simply a debate about particular platforms and specific requirements, it is fundamentally a debate about long-term strategy – about how the United States retains a hard won technological edge in the disruptive early phases of a robotics revolution that is changing the world.

---

[1] See Robert O. Work and Shawn Brimley, 20YY: Preparing for War in the Robotic Age (Washington DC: Center for a New American Security, January 2014).
[2] See Andrew Erikson and David Yang, "On the Verge of a Game Changer," Proceedings of the U.S. Naval Institute (May 2009). See also Gormley, Erickson, Yuan, A Low-Visibility Force Multiplier: Assessing China's Cruise Missile

CONGRESSIONAL
TESTIMONY

**Aim Higher: Elevating the Debate Over
Unmanned Carrier Aviation**
Prepared Statement of Shawn Brimley

Center for a
New American
Security

A Matter of Strategy

Requirements for military systems must flow from strategy. The primary strategic driver for U.S. power projection requirements is the evolving military competition in Asia. Since the end of World War II, the United States has acted as the ultimate guarantor of regional peace and security in Asia – sustaining a network of treaty allies and close partners to ensure a regional order commensurate with U.S. interests prevails. This order is straining under the weight of China's rise, manifesting in part by its increasingly bellicose and hostile behavior towards its neighbors, coupled with its major investments in modern military technology designed to exploit the vulnerabilities inherent in U.S. power projection strategy and capabilities. There are other actors that could pose quite plausible challenges to the U.S. Joint Force, but I believe China constitutes the clear "pacing threat" to which force planners must focus attention.

Given China's behavior, America's allies and partners in Asia are looking to us to maintain order. To do so the United States must maintain a credible conventional military deterrent in Asia. This requires that U.S. power projection capabilities and concepts can prevail across a range of plausible contingencies against China, any one of which requires the ability to penetrate their increasingly robust "anti-access / area-denial" (A2/AD) network, which includes advanced integrated air defense systems, fifth generation fighter aircraft, robust submarine forces, and increasingly precise long range anti-ship ballistic and cruise missiles.

All elements of China's A2/AD network are cause for concern, but it is their long range anti-ship ballistic missiles that most complicate naval airborne power projection. A good example is China's DF-21D missile, one that some analysts term a game-changing "carrier-killer" due to its ability to fly beyond the unrefueled range of a U.S. carrier's strike aircraft.[2] If the Navy fails to develop an answer for this Chinese weapon, the U.S. aircraft carrier may very well fade into irrelevance, as a commander-in-chief is unlikely to send in a Carrier Strike Group so easily targeted by an adversary's long-range missiles until the threat is significantly degraded. If the Navy needs to invent new weapons to eliminate the threat of long-range missiles, or if these missiles must be taken out before the carrier can come within an operationally meaningful distance from an adversary, it becomes harder to justify spending tens of billions of dollars on aircraft carriers at all. Given the value and prestige of aircraft carriers ($11 billion apiece for the

---

[2] See Andrew Erikson and David Yang, "On the Verge of a Game Changer," Proceedings of the U.S. Naval Institute (May 2009). See also Gormley, Erickson, Yuan, A Low-Visibility Force Multiplier: Assessing China's Cruise Missile Ambitions (Washington DC: National Defense University Press, 2014).

next-generation Ford-class) coupled with their planned 50-year lifespan, keeping them relevant for the conflicts of the future is critical.[3]

The primary way to keep the U.S. aircraft carrier relevant to future conflicts is to ensure that its embarked air wing has, in aggregate, sufficient stealth and strike power to penetrate adversary airspace and find and engage the full range of target types (fixed, mobile, re-locatable) – and do so while the carrier stays at a safe operating distance from the worst of an adversary's anti-ship ballistic and cruise missiles inventory. Thus we need the ability to operate stealthy strike aircraft at very long ranges from U.S. aircraft carriers.

The need to operate carrier strike aircraft at longer ranges is not a new development. The Defense Department's 2006 Quadrennial Defense Review required the Navy to "develop an unmanned longer-range carrier-based aircraft capable of being air-refueled to provide greater standoff capability, to expand payload and launch options, and to increase naval reach and persistence."[4] The case was reinforced in a 2008 paper by (now) Deputy Defense Secretary Robert Work and Thomas Ehrhard, who argued that an unmanned, carrier-based strike system was the only way to "provide the Navy's future carrier air wings with the organic, extended-range, survivable, and persistent surveillance-strike capability needed to meet a number of emerging 21st century security challenges."[5]

Prioritizing the introduction of unmanned strike aircraft into the carrier air wing is necessary not only to keep the carrier relevant in plausible high-end contingencies, but to unlock the benefits of unmanned technology. The ability to deploy unmanned systems with air-to-air refueling capability would dramatically extend the endurance of the air wing, making possible 30 to 40 hours of continuous operations compared to the 10 to 14 hours for a manned aircraft.[6]

A key priority for defense planners must be to ensure U.S. aircraft carriers can project power from the sea onto land anywhere in the world. In order to perform this mission in the future, the carrier air wing must be able to strike effectively over ranges much larger than the radius of an adversary's anti-ship missiles. If we can't do this, the nation's aircraft carriers, and the hundreds

---

[3] This argument is more fully explored in Shawn Brimley, "Congress's Chance to Fix Aircraft Carrier Drones," *Defense One* (May 4, 2014). A more ambitious argument favoring moving away from large-deck aircraft carriers altogether can be found in Henry J. Hendrix, At What Cost a Carrier? (Washington DC: Center for a New American Security, March 2013).

[4] Department of Defense, Quadrennial Defense Review (Washington DC, 2006): p. 46.

[5] Robert Work and Thomas Ehrhard, Range, Persistence, Stealth, and Networking: The Case for a Carrier-Based Unmanned Combat Air System, (Washington DC: Center for Strategic and Budgetary Assessments, 2008): p.7.

[6] This idea is explored in greater depth in Work & Ehrhard and also Paul Scharre, Robotics on the Battlefield (Washington DC: Center for a New American Security, 2014): p.19-22.

of billions they have cost to procure and operate, will likely fade into irrelevancy—perhaps sooner rather than later. That is a completely unacceptable outcome given the centrality of maritime power projection to U.S. national security strategy.

A Matter of Mission

The reason advocates of a more ambitious carrier-based unmanned aircraft system have been speaking out in recent months is due to indications that the Navy is not taking the imperative to ensure carrier-based long-range strike all that seriously.[7] It is a mission that must be fully resourced if carriers are to remain the nation's capitol ship and a major pillar of our forward deployed conventional deterrent.

There have been a series of directions given to the U.S. Navy articulating priorities for carrier-based unmanned systems. As recently as January 2014, Secretary of the Navy Ray Mabus wrote:

> "The end state [for UCLASS] is an autonomous aircraft capable of precision strike in a contested environment, and it is expected to grow and expand its missions so that it is capable of extended range intelligence, surveillance and reconnaissance (ISR), electronic warfare, tanking, and maritime domain awareness. It will be a warfighting machine that complements and enhances the capabilities already resident in our carrier strike groups."[8]

While elements of the recent draft request for proposals (RFP) for UCLASS are classified, reporting indicates that the RFP essentially biased toward establishing an unmanned ISR aircraft for the carrier, albeit one with a light strike capability and a moderate level of stealth capability (although reportedly not enough stealth for truly penetrating, persistent surveillance-strike operations).[9] I believe that the UCLASS program ought to be geared more toward being a full-spectrum "warfighting machine," with enough stealth and strike power to be able to, as the 2012 Defense Strategic Guidance articulated with respect to the Joint Force, "operate effectively in

---

[7] See Mark Gunzinger and Bryan Clark, "The Next Carrier Air Wing," Defense News (February 24, 2014), Paul Scharre, "Is the DOD Innovating? How the New Budget Stacks Up," War on the Rocks (March 17, 2014), Shawn Brimley, "Congress' Chance to Fix Aircraft Carrier Drones," Defense One (May 4, 2014), Bryan McGrath, "HASC Send Strong Message on UCLASS," Information Dissemination (April 30, 2014).

[8] Ray Mabus, "Future Platforms: Unmanned Naval Operations," War on the Rocks (January 21, 2014). Emphasis added. The key term here is "contested environment."

[9] See Dave Majumdar, "Navy Issues Restricted UCLASS Draft Request for Proposal," USNI News (April 17, 2014). Also see Sam LaGrone, "Navy Docs Reveal UCLASS Minimum Ranges and Maximum Costs," USNI News (June 26, 2013).

CONGRESSIONAL TESTIMONY

Aim Higher: Elevating the Debate Over Unmanned Carrier Aviation
Prepared Statement of Shawn Brimley

Center for a
New American
Security

anti-access and area denial (A2/AD) environments.[10] In this respect the actions of this committee to hold funding for UCLASS until the Secretary of Defense certifies requirements were justified.

Charitably, there are probably those who believe that the best way to integrate unmanned systems onto aircraft carriers is to make the transition as least disruptive as possible – both in terms of competition for missions, cost, and integration. I think it is fair to suggest that integrating an unmanned system designed for strike operations would be more difficult than one designed primarily for organic ISR on a 12-hour surveillance cycle.

But prioritizing the mission of organic ISR support to carrier operations seems odd, given the ability of other Navy platforms to perform this mission, including the P-8 Poseidon, the MQ-4C Triton, the MQ-8C Fire Scout, and the E-2D Advanced Hawkeye. It may be that an unmanned carrier-based ISR platform could provide some additional surveillance capability into this mix, but weighed against the primary mission of projecting meaningful offensive striking power against the nation's adversaries, using UCLASS to add more ISR for the carrier represents a significant missed opportunity.

It is also possible that those steering the requirements process inside the Pentagon felt that the Navy needed to better ensure that the aircraft carrier could help contribute to filling a robust demand for counterterrorism strike missions given the gradual drawdown in Afghanistan.[11] But parking an aircraft carrier, the capital ship of the U.S. Navy, off the coast of places like Yemen or Pakistan to deploy unmanned aircraft in a light strike counterterrorism role in uncontested or lightly contested airspace is probably not an optimal solution given the need to maintain a robust conventional deterrent and power projection capability for plausible high-end contingencies in Asia or the Middle East. Such a view would be the product of absolute worst-case assessments of future access-agreements for land-based unmanned systems—a scenario that seems implausible given the breadth of current access in the Middle East and Central Asia, and the depth of longer-ranged land-based options for unmanned aircraft at U.S. disposal.

A Matter of Timing

As described above, there are essentially two competing ideas for UCLASS: a semi-stealthy aircraft with sufficient endurance to operate off-cycle with normal carrier air wing operations to provide intelligence, surveillance, and reconnaissance (ISR) and light strike in lightly contested

---

[10] See the January 2012 Defense Strategic Guidance, which directed DOD to "invest as required to ensure its ability to operate effectively in anti-access and area denial (A2/AD) environments."

[11] See Dave Majumdar and Sam LaGrone, "House Committee Seeks to Stall UCLASS Program Pending New Pentagon Unmanned Aviation Study," USNI News (April 29, 2014).

78

CONGRESSIONAL TESTIMONY

Aim Higher: Elevating the Debate Over Unmanned Carrier Aviation
Prepared Statement of Shawn Brimley

Center for a
New American
Security

environments; and a more capable aircraft with air-to-air refueling capability designed to operate in denied airspace for penetrating surveillance and strike missions.

It is probably true that a less ambitious ISR program that doesn't require true integration into the manned air wing would be less expensive and less disruptive to traditional concepts of operation. It might even be possible to field a less capable system earlier, though the recent successful tests of the X-47B—an operational class prototype of a stealthy, air-refuelable, large-payload UAS—would seem to undermine this assertion.[12]

But taking several years to procure a carrier-based unmanned ISR system before undertaking a more ambitious strike system would essentially preclude any real development of an unmanned combat aircraft system (UCAS) for at least a decade. Given the pace of technological diffusion and the rapidity of China's military modernization, waiting a decade before fielding a system that can enhance the striking power of U.S. aircraft carriers seems particularly unwise—especially when all of the Navy's carrier-based unmanned aircraft developmental efforts to date have been aimed at reducing technical risk on just this class of system.

Rather than wait until the late-2020s to introduce unmanned strike aircraft into the carrier air wing, DOD should tackle this ambitious challenge now. We simply do not have time to wait. It is not as though advocates of procuring a more capable system are embracing a fantasy divorced from what is technically possible—the successful testing of the X-47B proves that the United States is on the cusp of achieving this kind of capability. To essentially turn away from the promise of a real carrier-based unmanned combat aircraft and go down a decade-long UCLASS cul-de-sac comes close to defense strategy malpractice.

A Matter of Innovation

The history of military innovation is also one of military culture, traditions, and legacy. These are powerful and important features of modern militaries and should not be discounted or necessarily dismissed outright. But civilian policymakers in the executive and congressional branches should be cognizant that sometimes new technologies or innovative concepts of operation that might threaten traditional approaches produce antibodies that can stymie innovation and ultimately pose the very real risk that tomorrow's military will not be in a position to fight and win the nation's wars. Consider if elements of the U.S. Army had been successful in preventing the adoption of the tank in favor of horses? What would have been the result if elements in the U.S. Navy had been successful in preventing the adoption of steam-

---

[12] The X-47B not only demonstrated that it is possible to launch and recover an unmanned carrier-based aircraft, but also that a tailless flying wing with stealthy characteristics could operate from a carrier.

powered ships? How might history have evolved if the U.S. Army had been successful in resisting the emergence of air power?[13] We stand today at a similar strategic crossroads in the introduction of unmanned systems into the Joint Force. To pass up this opportunity by aiming for a less capable unmanned carrier aircraft when the technology exists for something more capable and strategically relevant is to put future American military dominance at risk.

One way to interpret the obstacles that have been consistently placed in front of the various unmanned programs over the past decade is a general resistance to the prospect of unmanned aircraft gradually crowding out the traditional role of manned strike aircraft. This concern is overblown, as the carrier air wing will continue to be dominated by manned aircraft for the foreseeable future. What is more likely over the mid- to long-term is manned and unmanned aircraft will work together and manned-unmanned teams will evolve that employ any number of creative concepts of operation that leverage the unique abilities of both.[14] We see early indications of this kind of approach in how the U.S. Navy is planning to operate the manned P-8 Poseidon and the unmanned MQ-4C Triton.

Unmanned Carrier Aviation as a Pillar of a 21st Century Offset Strategy

Military culture can sometimes stifle innovation, but the U.S. military has also risen to the challenge many times before. Perhaps the most successful military-technical defense strategy ever developed by the United States was the "offset strategy" developed in the latter decades of the Cold War.

During the late 1970s, Pentagon strategists struggled to maintain plausible conventional deterrence against a Soviet force that enjoyed a massive numerical advantage. Secretary of Defense Harold Brown and his then-Under Secretary of Defense William Perry developed a new technology strategy to address this dilemma. As Secretary Perry describes it, they faced the need to "develop high-technology systems that could give our military forces a qualitative advantage able to offset the quantitative advantage of the Soviet forces. Not surprisingly, this approach was called the 'Offset Strategy'."[15]

---

[13] See Williamson Murray and Allan Millet (eds), Military Innovation in the Interwar Period (New York: Cambridge University Press, 1996), and Carl Builder, The Masks of War: American Military Styles in Strategy and Analysis (Baltimore: Johns Hopkins University Press, 1989).

[14] For more on this see Paul Scharre, Robotics on the Battlefield: Range, Persistence and Daring (Washington DC: Center for a New American Security, 2014).

[15] William J. Perry, "Technology and National Security: Risks and Responsibilities," Speech to France-Stanford Center for Interdisciplinary Studies, April 7, 2003. Also see Ashton Carter, "Keeping the Technical Edge," in Carter and John White (eds), Keeping the Edge: Managing Defense for the Future (Cambridge, MA: Harvard University Press, September 2000).

The strategy centered on investing in several disruptive technologies that today seem prescient but at the time were anything but certain. By investing in stealth aircraft, precision weapons, advanced satellites, computer networking and other technologies, we developed the ability to coordinate precision strikes over long distances and, by doing so, undermined the Soviet military's numerical advantage. This was a major reason why Moscow was forced to nearly bankrupt itself in the attempt to develop countermeasures. Quintessentially asymmetric, the "offset strategy"— the concepts of operations and technologies developed—constitutes perhaps the most impressive defense investment strategy ever developed by the United States.

The United States has essentially been dining out on the Cold War offset strategy for a quarter-century—from the 1991 Gulf War to the present. Given the pace and scale of the military-technical challenges developing in the international system, the Department of Defense must get more ambitious and aggressive in ensuring that America's military-technical dominance persists.

We should see the debate regarding unmanned carrier-based aircraft in this strategic light—as a core feature in what we might term a new "offset strategy" that would include unmanned and increasingly autonomous systems; directed energy and electric weapons; robust cyberwarfare capabilities; advanced protected communications; and other game-changing systems. These constitute emerging defense investment 'vectors' in which the United States must lead, not follow, both to ensure a first-mover technical advantage, but learn to field and employ these technologies in operationally meaningful ways.[16]

Congress must ensure that DOD sustains its military technological dominance, as our military competitors will not hesitate to overtake us. A leading indicator of the seriousness with which the United States approaches this strategic imperative is the shape of the Navy's UCLASS program. It is a debate in which the more ambitious and more aggressive approach is the right one.

The stakes are high. If the United States fields a carrier-based unmanned combat air system within the next decade, it will go a very long way toward ensuring that tomorrow's adversaries fear the U.S. aircraft carrier and the long-range combat strike power it can unleash, and it will set the Department of Defense on the right path toward securing America's military-technical dominance for the next generation. Congress must ensure the Department of Defense aims high enough to meet these two strategic imperatives.

---

[16] For a more detailed exploration of what kinds of future military capabilities could underwrite a new "offset strategy" to ensure U.S. military-technical dominance, see Robert O. Work and Shawn Brimley, 20YY: Preparing for War in the Robotic Age (Washington DC: Center for a New American Security, January 2014).

Shawn Brimley
Executive Vice President and Director of Studies
Center for a New American Security

Shawn Brimley is Executive Vice President and Director of Studies at the Center for a New American Security (CNAS). Mr. Brimley served in the Obama Administration from February 2009 to October 2012 most recently as Director for Strategic Planning on the National Security Council staff at the White House. He also served as Special Advisor to the Under Secretary of Defense for Policy at the Pentagon from 2009 to 2011, where he focused on the 2010 Quadrennial Defense Review, overseas basing and posture, and long-range strategy development. He has been awarded the Secretary of Defense Medal for Outstanding Public Service and the Office of the Secretary of Defense Medal for Exceptional Public Service. Mr. Brimley was a founding member of CNAS in 2007 and was the inaugural recipient of the 1Lt. Andrew Bacevich Jr. Memorial Fellowship. He has also worked at the Center for Strategic and International Studies.

Mr. Brimley has published widely, including in the *New York Times, Foreign Affairs* and *Foreign Policy*. Educated at Queen's University and George Washington University, Mr. Brimley is a term member of the Council on Foreign Relations. He lives in Washington with his wife and their three children.

Shawn Brimley

EMPLOYMENT

Center for a New American Security   2013 to present
Executive Vice President and Director of Studies
- Member of executive team for bipartisan national security think-tank. Manages research staff and programs across life cycle—fundraising, execution, publication, and outreach (30 employees; ~$7 million budget).
- Helped implement new business model spurring increase in revenue, cash reserve, and research output. Significantly increased percentage of women and minorities across research staff.

The White House, National Security Council   2011 to 2012
Director for Strategic Planning
- Supported policy planning and coordination on national security priorities as well as crisis management.
- Focus areas included South and Southeast Asia, defense planning and posture, and budget processes.

U.S. Department of Defense   2009 to 2011
Special Advisor to the Under Secretary of Defense for Policy
- Served as advisor to Under Secretary Michele Flournoy on defense strategy and planning issues.
- Lead writer for 2010 Quadrennial Defense Review. Involved in cyber, nuclear and force posture reviews.

Center for a New American Security   2007 to 2009
Fellow
- Inaugural recipient of the 1Lt. Andrew J. Bacevich, Jr., USA Memorial Fellowship.
- Work focused on Iraq and Afghanistan, defense policy, and national security strategy.

Center for Strategic and International Studies   2005 to 2007
Research Associate
- Supported Senior Advisor Michèle Flournoy on a wide-ranging research and writing agenda.
- Work focused on defense policy, interagency reform, counterterrorism, and counterproliferation.

EDUCATION
MA in Security Policy Studies, George Washington University, Elliott School of International Affairs
BA in History, Queen's University, Kingston, Ontario

SELECT PUBLICATIONS

"Resetting the U.S. Military," with Paul Scharre, *Foreign Policy* (May/June 2014).
20YY: Preparing for War in the Robotic Age, with Robert Work (CNAS, January 2014).
"The Drone War Comes to Asia," with Ely Ratner and Ben FitzGerald, *Foreign Policy* (September 2013).
"Smart Shift: A Response to 'The Problem with the Pivot'," with Ely Ratner, *Foreign Affairs* (January 2013).
"The Defense Inheritance," with Michèle Flournoy, *The Washington Quarterly* (Fall 2008).
"How to Exit Iraq," with Colin Kahl and John Nagl, *The New York Times* (5 September 2008).
"The Endgame in Iraq," with Kurt Campbell, *Foreign Policy* (July/August 2007).
Phased Transition: A Responsible Way Forward and Out of Iraq, with James Miller (CNAS, June 2007).
Strategic Planning for U.S. National Security, with Michèle Flournoy, (Princeton University, August 2005).

OTHER EXPERIENCES
- Council on Foreign Relations term member
- Member of Defense and Iraq policy teams for "Obama for America" 2008 campaign
- Lived in Asia for several years (taught English in Japan; Habitat for Humanity in India and Philippines)

83

**DISCLOSURE FORM FOR WITNESSES
CONCERNING FEDERAL CONTRACT AND GRANT INFORMATION**

INSTRUCTION TO WITNESSES: Rule 11, clause 2(g)(5), of the Rules of the U.S. House of Representatives for the 113th Congress requires nongovernmental witnesses appearing before House committees to include in their written statements a curriculum vitae and a disclosure of the amount and source of any federal contracts or grants (including subcontracts and subgrants) received during the current and two previous fiscal years either by the witness or by an entity represented by the witness. This form is intended to assist witnesses appearing before the House Committee on Armed Services in complying with the House rule. Please note that a copy of these statements, with appropriate redactions to protect the witness's personal privacy (including home address and phone number) will be made publicly available in electronic form not later than one day after the witness's appearance before the committee.

Witness name: _____Shawn Brimley_____

Capacity in which appearing: (check one)

__X_Individual

____Representative

If appearing in a representative capacity, name of the company, association or other entity being represented:

FISCAL YEAR 2014

federal grant(s)/ contracts	federal agency	dollar value	subject(s) of contract or grant
NONE			

FISCAL YEAR 2013

federal grant(s)/ contracts	federal agency	dollar value	subject(s) of contract or grant
NONE			

84

FISCAL YEAR 2012

Federal grant(s) / contracts	federal agency	dollar value	subject(s) of contract or grant
NONE			

Federal Contract Information: If you or the entity you represent before the Committee on Armed Services has contracts (including subcontracts) with the federal government, please provide the following information:

Number of contracts (including subcontracts) with the federal government:

Current fiscal year (2014):_____NONE_____;
Fiscal year 2013:_____;
Fiscal year 2012:_____.

Federal agencies with which federal contracts are held:

Current fiscal year (2014):_____NONE_____;
Fiscal year 2013:_____;
Fiscal year 2012:_____.

List of subjects of federal contract(s) (for example, ship construction, aircraft parts manufacturing, software design, force structure consultant, architecture & engineering services, etc.):

Current fiscal year (2014):_____NONE_____;
Fiscal year 2013:_____;
Fiscal year 2012:_____.

Aggregate dollar value of federal contracts held:

Current fiscal year (2014):_____NONE_____;
Fiscal year 2013:_____;
Fiscal year 2012:_____.

Federal Grant Information: If you or the entity you represent before the Committee on Armed Services has grants (including subgrants) with the federal government, please provide the following information:

Number of grants (including subgrants) with the federal government:

        Current fiscal year (2014):_____NONE_____;
        Fiscal year 2013:_____;
        Fiscal year 2012:_____.

Federal agencies with which federal grants are held:

        Current fiscal year (2014):_____NONE_____;
        Fiscal year 2013:_____;
        Fiscal year 2012:_____.

List of subjects of federal grants(s) (for example, materials research, sociological study, software design, etc.):

        Current fiscal year (2014):_____NONE_____;
        Fiscal year 2013:_____;
        Fiscal year 2012:_____.

Aggregate dollar value of federal grants held:

        Current fiscal year (2014):_____NONE_____;
        Fiscal year 2013:_____;
        Fiscal year 2012:_____.

Testimony before the House Armed Services Committee
Subcommittee on Seapower and Projection Forces

Prepared Statement of Bryan McGrath
Managing Director, The FerryBridge Group LLC and Assistant Director, Hudson
Institute Center for American Seapower

July 16, 2014

All testimony herein represents the personal views of Bryan McGrath

Thank you Chairman Forbes and Ranking Member McIntyre and all the members of the Seapower and Projection Forces subcommittee for the opportunity to testify and to submit this written statement for the record.

I am a defense consultant by trade, specializing in naval strategy. Additionally, I recently joined with Seth Cropsey of the Hudson Institute to found a think tank devoted to Seapower, known as the Hudson Center for American Seapower. All of my adult life has been spent either in the Navy or working on matters of naval operations and strategy.

On active duty, I commanded a destroyer, and I was the team leader and primary author of the 2007 USN/USMC/USCG maritime strategy known as "A Cooperative Strategy for 21st Century Seapower. Since leaving active duty in 2008, I have written and spoken widely about preponderant American Seapower as the element of our military power most that most effectively and efficiently promotes and sustains America's prosperity, security, and role as a world leader.

I am concerned that there is insufficient understanding among the American people and its leaders of the relationship between preponderant Seapower and our national greatness. I am concerned that there is insufficient attention being paid by the American people and many of its leaders to the dramatic and potentially irreversible impact of recent budget cuts on American Seapower. I am concerned that the Obama Administration has not backed up the strategic aspirations embodied in its "rebalance" to the Pacific with rational resource planning and tough strategic choices. Mostly, I am concerned about rising Chinese power and the threat it poses to the global order from which we in the United States benefit greatly (and truth be told, for which we pay disproportionately).

Others on this panel will eloquently describe the nature of the Chinese threat. There has been no shortage of discussion in the various defense journals of "Anti-Access/Area Denial Threats" and the clear desire of the People's Liberation Army (PLA) to execute a counter-intervention strategy that seeks to deny the United States the ability to project significant military power. Chinese modernization trends clearly stress the desire to create conditions under which our access and influence in the region are diminished. While some focus on the means of this strategy, I would emphasize the ends, which are to undermine our alliance system in the Asia Pacific region. The United States MUST contest this strategy. It must not cede significant portions of the earth's surface because other powers develop weapons that increase risk to our forces. We must lean forward by capitalizing on one of our most important competitive advantages, by which I mean our research and development base and our ability to change strategic conversations with the power of our ideas and the output of our industrial base.

Which brings us to the subject of today's hearing, the Navy's Unmanned Carrier Launched Surveillance and Strike System, hereafter referred to as UCLASS. Much has been made in the press about the Navy's plans for this system and, because the exact requirements remain justifiably classified, we cannot know with certainty the direction the acquisition will take.

But there have been troubling reports that lead me to believe that the specifications for the system vastly over-privilege surveillance at the cost of capable strike in contested electromagnetic and surface-to-air missile environments. I believe some members of this Sub-committee agree with me. This is not a small point. As a matter of fact, it is a very significant one. It is potentially the beginning of "the beginning of the end" of America's preponderance at sea.

The centerpiece of America's forward deployed power projection capability is the aircraft carrier strike group, or CSG. The CSG should be thought of as a sea-based combat system for the command and control of battle-space (specifically, the seas and skies in which it operates) and the projection of power ashore. This largely self-sufficient unit of American military power has contributed to the proper functioning of the global system of trade and finance for over sixty years by ensuring that no nation has the capability threaten the freedom of the sea commons. At the heart of this capability is the nuclear powered aircraft carrier, an instrument of remarkable flexibility and adaptability that has deterred conflict and assured allies and friends for decades, even as critics routinely (re)raise notions of its obsolescence. It is the ultimate expression of our interest in the region and our ability to influence friends.

The secret to the aircraft carrier's centrality in American defense planning has been the simple fact that, to a large extent, the platform is agnostic to the weapons it wields. No other element of America's arsenal has so thoroughly adapted with the times, as aerospace technologies provided for ever increasing capability. Over the years, the "main battery" of the aircraft carrier—known as the "Carrier Air Wing" (henceforth, CVW)—has continuously evolved to ensure the U.S. Navy operates with a comfortable margin of superiority over all potential adversaries.

History has not reached its end and, for the U.S. Navy to continue to exercise preponderant Seapower in the 21st Century, the CVW must continue to evolve to reflect the state of technology and the viability of the threat. The importance of CVW evolution is precisely why Secretary of the Navy Ray Mabus' early prioritization of unmanned surveillance and strike capability on a U.S. carrier was such a clear and powerful statement of purpose, notable from an Administration that has thus far not shown a grasp of the importance of American Seapower.

Simply put, if America does not devote itself to fielding unmanned, autonomous strike platforms capable of operating in contested environments, we may in fact reach the point where six decades of predictions finally come true—that the aircraft carrier will have reached its point of obsolescence, taking with it billions of dollars of taxpayer investment and prudent operational planning, simply because we did not have the courage and foresight to field the capability required to sustain our Navy's ability to operate where it matters, when it matters. Put another way, if we do not insist that the Navy put the UCLASS acquisition on a path to creating a future air wing of manned and unmanned platforms capable of operating as a system to counter adversary strategies in contested environments, we are likely to see the dominance of the American Navy wane and, with it, the network of alliances and friendships that has underpinned American security and global prosperity for decades.

Why Seapower? American Seapower is the most flexible of the various instruments of military power, and the one uniquely able to accommodate our desire for a peaceful and prosperous world. Even more, it is an essential element of an effective grand strategy, along with a strong economy and useful alliances. As policy-makers begin to think seriously about an appropriate grand strategy for the Post War on Terror world, American Seapower should occupy a central position. Several obvious US national security imperatives are made possible by preponderant American Seapower.

Seapower Enables the Homeland Defense "Away Game". Naval forces operate for extended periods far from US shores without the permission of any sovereign government; this translates into the extension of America's homeland "defensive perimeter". The ability to gather information, perform surveillance of seaborne and airborne threats, interdict suspected WMD carriers and disrupt terrorist networks without a large shoreward "footprint" is critical in a world of denied access and decreasing acceptance of American troops stationed abroad. Dealing with these threats as far from our shores as possible gains decision space and time for political and engagement opportunities.

Seapower Bolsters Critical Security Balances. Preponderant American Seapower underwrites East Asian security by demonstrating to Allies and friends American resolve to maintain regional stability. Additionally, the overwhelming advantage enjoyed by US naval forces in sea control and striking power is, in and of itself, an inducement to maintaining security. Absent such preponderance, a nascent Asian naval arms race has the potential to intensify, with predictably deleterious effects for the United States and our Allies. In the Arabian Gulf and Indian Ocean, sustained preponderant US naval combat power serves to assure allies of the nation's resolve to maintain stability in the face of an unpredictable regime in Iran.

Seapower Provides an Effective Conventional Deterrent. The visible presence of American Seapower operating freely in the maritime commons provides an effective conventional deterrent to those who would seek to threaten regional security and stability. First, the capabilities and capacities of preponderant naval power are arrayed in a manner that causes an adversary to question the effectiveness of a pre-emptive attack (deterrence by denial). Such capabilities include sea-based ballistic missile defense (BMD) and the striking power of carrier-based airpower armed with precision guided munitions. Second, the likelihood of a prompt and painful counter-attack from the sea raises the costs associated with military adventurism (punishment). In either case, recent scholarship in the study of conventional deterrence indicates that *overall* US conventional superiority is less likely to provide an effective deterrence than is the *local regional balance of power.* This suggests that in order to deter effectively, the US must be "present" – and no form of military power can be as consistently present in as many critical places at once as Seapower.

Seapower Enables Diplomacy, Development and Defense. American Seapower is the global guarantor of freedom of commerce on the world's oceans, thereby promoting American economic stability and growth. This role has been played before in history by the Portuguese, the Dutch and the British, but never before has it been played by a nation without imperial or colonial aspirations. American guarantees to the global commons do not come with a colonial "tax" on other nations. The overwhelming majority of world trade (by weight and by value) travels across the world's oceans, to the benefit of all trading nations. Additionally, America's diplomatic power is increasingly enabled by its Seapower, a symbiotic relationship reminiscent of US foreign policy conduct throughout much of its pre-World War II history. American Seapower provides its statesmen and diplomats with new options for flexibly engaging Allies, partners, friends and others. This close relationship between America's naval forces and its diplomatic arm will be essential to promoting good governance in ungoverned spaces and building partnership capacity in nations facing critical security threats.

Seapower Provides for Modulated Military Response. The world is an increasingly disordered and untidy place, with regional instability a constant feature of the strategic landscape. Should deterrence fail (as it sometimes does), already present, combat ready naval forces are prepared to conduct prompt and sustained operations. These operations range from shows of force, raids and demonstrations, strikes and special operations, all the way to the forcible entry of land power from the sea. This menu of choices is a primary feature of American Seapower, and it provides the President with unmatched flexibility to respond, escalate, and de-escalate without having to deploy additional forces from the United States. Should the nation find it necessary to transition to a punishing land war, American Seapower provides the means for assuring the entry of follow-on forces, as well as providing considerable combat power in support of ongoing land operations.

Seapower Provides America's Survivable Nuclear Deterrent. The Navy's fleet of 14 ballistic missile submarines (SSBN) -- each equipped with Multiple Independent Re-entry Vehicle (MIRV) armed Trident Submarine Launched Ballistic Missiles (SLBMs) — is our most survivable method of providing strategic nuclear deterrence. With Russia increasing its reliance on nuclear weapons and China upgrading its own nuclear stockpile — in addition to the nuclear mischief of North Korea and Iran — the US must continue to upgrade its SSBN force even as it considers new and novel ways to employ them.

Seapower Shows the Best Face of America. The purpose of American military power is to protect the United States by fighting and winning wars, and American Seapower is no exception. That said, the staggering cost of military power demands a premium be placed on those forces with *peacetime missions* that *also* advance the national security of the United States. No nation on earth is as quick to provide humanitarian assistance in the wake of natural and humanitarian disasters as the US, and no element of American power is as critical to prompt and sustained recovery efforts as American Seapower. Whether it is the direct provision of food, water and shelter; emergency medical care; or security in a chaotic environment, it is American Seapower that answers the nation's call when its considerable sympathy moves it to act.

Why is UCLASS Critical to Sustaining American Seapower?

When the Berlin Wall fell, the United States was left as the sole superpower, able to project power from the sea wherever it needed without serious fear either of opposition or reprisal. America's Navy was ascendant, and its ability to control the seas (the necessary precondition to project power from the sea) was unquestioned. As a result, the CVW evolved from its Cold-War era instantiation which included longer range strike assets and sea control aircraft (both of which were necessary to contest a near peer in the Soviet Navy) to one which featured much shorter range strike aircraft capable of higher sortie rates, leveraging the precision guided weapons revolution. Sea control aircraft—which prosecuted enemy surface ships and submarines—were largely removed from the air wing, with the strike/fighter squadrons assuming greater sea control responsibilities.

This arrangement was sufficient so long as no one contested America's ability to control the seas. That salutary condition began to wane in the 21st century, as China worked to assemble a family of capabilities designed to ensure that American forces would not be capable of operating close enough to its shores (or more to the point, to Taiwan's shores) for it to be able to generate the massive amount of power projection necessary to achieve major military objectives. Keep the Americans outside the combat range of their power projection platforms, and America loses its competitive advantage. Chief among the strategies for accomplishing this goal was the development of a series of "anti-access/area denial" measures designed specifically to target the aircraft carrier, which the PLA rightly identified as the lynchpin of American forward combat power.

It must be remembered that the U.S. Navy made a conscious decision to alter the makeup of its carrier air wings after the dissolution of the Soviet Union, trading range for sortie generation, a luxury afforded it by a lack of any real threat. Again, the nuclear-powered aircraft carrier itself did not fundamentally change; the weapons system it projected did. China has created a series of weapons (missiles -- both ballistic and cruise -- long range bombers and submarines) designed to increase the threat to the carriers which (by their logic) would cause us to operate them well-outside the effective combat range of their air wings. If our air wings do not evolve to once again "buy back combat range," then the Chinese strategy will have succeeded.

In addition to a whole host of countermeasures designed to attack China's ability to find, fix, target the aircraft carrier in the first place, the Navy must evolve its air wings to conduct strike operations at longer ranges in non-permissive environments. When UCLASS was first announced, it was assumed by many that it would be the first step in fielding an unmanned capability that would operate side by side with manned aircraft on the carrier decks to accomplish this goal. As the Joint Strike Fighter begins to populate carrier decks, providing fifth generation capabilities and increased strike range, UCLASS would evolve to eventually replace the F/A-18 E/F Super Hornet fleet with an unmanned, autonomous platform capable of operations in a contested environment, including complex electromagnetic and air defense threats. This long-range vision of manned and unmanned strike vehicles extending the useful operational range of the aircraft carrier dramatically above its current striking distance reinforced the centrality of the aircraft carrier in America's forward deployed power projection scheme.

This ability to generate combat power at greater distances directly counters China's A2AD regime, and serves as a defining capability to what has come to be called "Air Sea Battle". Much has been made of Air Sea Battle in the open press, and much of that has been overheated and wrongheaded. What has received insufficient attention in the Air Sea Battle debate is the deterrent value gained by ensuring the PLA knows that its A2AD regime can be effectively countered. A carrier air wing capable of striking targets from outside the likely operational range of China's A2AD complex provides a powerful incentive to Chinese leaders NOT to incite conflict that they know would bring a swift and powerful reprisal. Additionally, a U.S. capability to effectively counter China's A2AD complex provides assurance to our Allies and friends in the region that we will not be ejected from the Western Pacific, removing the temptation for them to pursue separate accommodations with the Chinese.

Ensuring that UCLASS requirements account for appropriate levels of stealth, autonomy, range and lethality that enable it to operate in contested environments will ensure that the carrier air wing continues to evolve in a manner that leverages the mobility and flexibility of the nuclear powered aircraft carrier. This is critical to sustaining American Seapower, which I believe is critical to sustaining our security, prosperity, and global leadership.

What is Wrong with a UCLASS that Privileges Surveillance and Precludes Strike?

The primary problem with a UCLASS that overly privileges surveillance at the cost of strike operations in a contested environment is that, in a time of tight budgets, we can ill afford to build and field yet another ISR system that cannot penetrate and attack. The Navy is already purchasing 68 MQ-4C Triton UAV's and in excess of 100 P-8 manned MPRA aircraft; therefore, a carrier based, non-stealthy UAV that stresses mission duration over stealth and strike is largely duplicative to these two more capable systems. That a non-stealthy ISR heavy UCLASS might have some limited capability for strike in a permissive environment will not justify even the first dollar that would be spent on it.

The opportunity costs associated with pursuing a surveillance-heavy UCLASS are immense, not the least of which could be the potential for realizing the six-decade old predictions of the end of the aircraft carrier. It is not my intention to advocate for a stealthy, strike UCLASS in order to "save the aircraft carrier". It is my intention to advocate for a stealthy, strike UCLASS because I fear that if we do not move in this direction, the PLA anti-intervention strategy will largely succeed, and the preponderant Navy that we enjoy today will become a thing of the past. Unless and until something comes along that enables the United States to deter and assure from the sea with the success of the aircraft carrier and its embarked air wing, we must continue to evolve the air wing to ensure that the utility and flexibility of the carrier will continue to be manifest.

Why Would the Navy Proceed with a Surveillance-heavy UCLASS?

The simple answer is resources. Clearly, a non-stealthy, surveillance privileged platform with limited or zero autonomy and without the ability for in-flight refueling would represent considerably less acquisition risk than a stealthy, autonomous, in flight refuelable platform. Were cost the only (or even the main) consideration, such a path would be worth considering.

But acquisition cost is not the only — or even the main — consideration; or at least it should not be. The main consideration of this program should be to ensure a future path to manned and unmanned carrier air wings capable of operations within an enemy's desired "keep out" zone, thereby contributing to the continuation of preponderant American Seapower. Allocating considerable resources to a UCLASS that does not advance THIS goal may be less expensive in the short run, but it is MORE wasteful than pursuing the (admittedly) more challenging and expensive goal of stealthy, autonomous, strike in a contested environment

Critics of my view strike me as being in a hurry to get an admittedly limited system fielded as soon as possible, with the goal of making enhancements as the program progresses. This is often a path that I advocate, but in this instance, the significant differences in planform and associated propulsion options needed to support the more challenging strike missions strongly suggest that a surveillance privileged UCLASS simply could never evolve to meet more stringent, contested environment requirements. And so rather than move forward in a direction that largely duplicates existing Navy ISR capability while offering no enhancements to the carrier air wing versus a complex A2AD environment, I recommend continuing to work to ensure that the Navy's UCLASS requirements effectively addresses the operational problems posed by China's A2AD complex.

Additionally, I urge this Sub-committee to closely monitor the Navy's ongoing plans for its two X-47B UCAS platforms, the UAV that captured the country's imagination last summer by taking off and landing on an aircraft carrier autonomously. Reports in the July 7 "Inside the Navy" indicate that the Northrop Grumman program manager claims to be prepared to execute aerial refueling capability during upcoming shipboard testing aboard USS THEODORE ROOSEVELT (CVN 71). The contract under which this testing will occur reportedly contains an option to demonstrate autonomous in-flight refueling, and the Navy's unmanned carrier aviation program manager was quoted in the story as stating that such testing would proceed, "…if resources allow…" In-flight refueling will be key to extending the range of a stealthy, strike platform, and the Navy should move mountains to ensure that resources will allow such testing.

Finally, I urge this subcommittee to query the Secretary of the Navy as to why — if unmanned capability is such a high priority of his — the Navy has downgraded the position of Resource Sponsor for Unmanned Systems from a Rear Admiral to a Captain. Keep in mind, this person is the resource sponsor for the overwhelming number of unmanned systems in the Navy, to include undersea, aerial, and surface systems. Each of the major platform sponsors (air, surface, and submarines) remains a 2-star officer. This downgrade was made ostensibly in response to DoD-wide direction to cut flag billet numbers, but it clearly sends the wrong message and it seems antithetical to the Secretary's vision for advancing unmanned systems.

Bryan G. McGrath

CDR USN (ret.)
27414 Ferry Bridge Road
Easton, MD 21601
410.443.2187
bmcgrath@ferrybridgegroup.com
Cleared: TS/SI

WORK HISTORY

Managing Director, The FerryBridge Group LLC (2013-Present)
- *Founder of an independent consultancy focusing on National Security issues, Maritime Strategy, and Defense Technology development.*

Director of Consulting, Studies and Analysis, Delex Systems, Inc, Herndon, VA (2009-2013)
- *Founding Director of a consultancy focusing on Naval and National Security issues*

Manager, Strategic Planning, Northrop Grumman Marine Systems, Washington, DC (2008-9)
- *Primary Strategic Planner for a $500M line of business in commercial energy and defense.*

Director, Navy Strategic Actions Group. Washington DC (2006-8)
- *Senior Advisor to the uniformed leader of the US Navy (and member of the Joint Chiefs of Staff); responsible for formulating and implementing global strategy for the US Navy*

Commanding Officer, USS BULKELEY. Norfolk, VA (2004-2006)
- *CEO level position directing the activities of a $1 billion warship and crew of 320*
- *Air and Missile Defense Commander for Commander, IWO JIMA Expeditionary Strike Group*

Chief of Interoperability, Joint Staff, Washington DC (2001-2004)
- *Director level position coordinating missile defense oriented acquisition programs of the US Armed Services*

Executive Officer, USS PRINCETON. San Diego, CA (1999-2001)
- *COO level position managing the activities of a $1 billion warship and crew of 410*

Special Assistant to the Chief of Naval Operations. Washington DC (1997-1999)
- *Director level position as Communications Director and Speechwriter to the uniformed leader of the US Navy (and member of the Joint Chiefs of Staff)*

Junior Officer Naval Service (1987-1997)

RELEVANT EXPERIENCE

Chief Navy Strategist (2006/7)
- **Key Contribution.** Led the Washington based team of USN, USMC, and USCG officers who developed the nation's current Maritime Strategy "A Cooperative Strategy for 21st Century Seapower", and served as its primary author.
- **Managing Complexity.** Led a team of nearly 200 senior military officers, academics, and government officials in developing the United States Maritime Strategy, the plan for investing nearly $120 billion dollars annually for the next ten years.
  - Hand-picked by Navy leadership to manage this first comprehensive strategy development effort in 20 years.
- **Public Speaker.** Created and executed an extensive national advocacy and outreach program in support of the development of the National Maritime Strategy, including symposia, newspaper editorials, targeted media, and Congressional liaison.
- **Foresight.** Coordinated an in-depth alternative futures and strategic environment assessment process to support the development of the Maritime Strategy, creating a visionary look at the major trends in globalization, trade, finance, technology and labor now used as the standard for Department of Defense planning.

Command at Sea (2004-2006)
- **Proven Leader.** Received the 2006 "Zumwalt Award for Inspirational Leadership" from the Surface Navy Association.
- **Efficient.** In command of USS BULKELEY, managed over $20 million in resources with recognition for operating fiscal efficiency. Earned 2006 USS ARIZONA Trophy for "most combat ready ship" in the Navy
- **Operational.** Served as the Air and Missile Defense Commander for the IWO JIMA Expeditionary Strike Group, responsible for the seamless integration of the Strike Group into existing Joint Air Defense Networks and the creation of such networks where none previously existed.
- **Organizational Improvement.** Re-organized the management team in USS BULKELEY to reflect functional areas related to combat operations, rather than historic administrative alignment. This innovation created increased communication among the stake-holders and ultimately contributed to the ship's recognition as the most combat ready ship in the Navy.
- **Process Improvement.** Reduced maintenance and repair costs in USS BULKELEY by implementing an in-depth analysis of maintenance request procedures, resulting in 10% faster turn-around on high priority repairs with 50% fewer requests rejected for errors. Maintenance costs were maintained at 80% of the class average throughout command tenure.

Joint Staff Officer (2001-2004)
- **International Expertise.** Experience working with European, Middle Eastern, Asian and Latin American partners. Served as the primary Joint Staff representative to the international data link community, with deep expertise in Link 16, CEC and other missile defense oriented information and weapon systems.
    - Dynamic leadership and emphasis on personal excellence resulted in a 20% increase in retention of key Sailors and a 75% increase in personnel promotion rates.
- **Skilled Negotiator.** Excelled as primary agent of the Joint Chiefs of Staff for oversight of weapon system interoperability. Aided defense acquisition process by coordinating 25 separate programs (totaling over $15 billion) in implementing higher levels of Joint interoperability, resulting in greater combat efficiency at lower total cost to the taxpayer.

EDUCATION

MA, Political Science, *The Catholic University of America*, 1999
BA, History, *University of Virginia*, 1987
Navy Fellow, *Massachusetts Institute of Technology Foreign Policy Seminar XXI*, 2007
Graduate, *Naval War College*, 1999 (JPME Phase I)
JPME Phase II (2006)

MISCELLANEOUS

Adjunct Fellow, Hudson Institute and Assistant Director of the Hudson Center for American Seapower (2013-Present)
Navy Policy Team Lead, Romney for President (2011-2012)

**DISCLOSURE FORM FOR WITNESSES
CONCERNING FEDERAL CONTRACT AND GRANT INFORMATION**

INSTRUCTION TO WITNESSES: Rule 11, clause 2(g)(5), of the Rules of the U.S. House of Representatives for the 113th Congress requires nongovernmental witnesses appearing before House committees to include in their written statements a curriculum vitae and a disclosure of the amount and source of any federal contracts or grants (including subcontracts and subgrants) received during the current and two previous fiscal years either by the witness or by an entity represented by the witness. This form is intended to assist witnesses appearing before the House Committee on Armed Services in complying with the House rule. Please note that a copy of these statements, with appropriate redactions to protect the witness's personal privacy (including home address and phone number) will be made publicly available in electronic form not later than one day after the witness's appearance before the committee.

Witness name: Bryan McGrath

Capacity in which appearing: (check one)

X Individual

___Representative

If appearing in a representative capacity, name of the company, association or other entity being represented:

FISCAL YEAR 2014

federal grant(s) / contracts	federal agency	dollar value	subject(s) of contract or grant
N00178-04-D-4148	USN (OPNAV N96)	$70K	Strategic Comms and Planning

FISCAL YEAR 2013

federal grant(s) / contracts	federal agency	dollar value	subject(s) of contract or grant
N00178-04-D-4148	USN (OPNAV N96)	$6K	Strategic Comms and Planning

FISCAL YEAR 2012

Federal grant(s) / contracts	federal agency	dollar value	subject(s) of contract or grant
N00178-04-D-4119-NS17	USN (OPNAV N4/9)	$75K	USN Strategy and Alignment
GS-10F-0339K	US Army	$277K	US Army AMD in the Pacific

Federal Contract Information: If you or the entity you represent before the Committee on Armed Services has contracts (including subcontracts) with the federal government, please provide the following information:

Number of contracts (including subcontracts) with the federal government:

Current fiscal year (2014): 1 ;
Fiscal year 2013: 1 ;
Fiscal year 2012: 2 .

Federal agencies with which federal contracts are held:

Current fiscal year (2014): USN ;
Fiscal year 2013: USN ;
Fiscal year 2012: USN, US Army .

List of subjects of federal contract(s) (for example, ship construction, aircraft parts manufacturing, software design, force structure consultant, architecture & engineering services, etc.):

Current fiscal year (2014): Surface Warfare Programs ;
Fiscal year 2013: Surface Warfare Programs ;
Fiscal year 2012: US Navy Program, Strategy. US Army Air and Missile Def .

Aggregate dollar value of federal contracts held:

Current fiscal year (2014): $70K ;
Fiscal year 2013: $6K ;
Fiscal year 2012: $352K .

Federal Grant Information: If you or the entity you represent before the Committee on Armed Services has grants (including subgrants) with the federal government, please provide the following information: No Grants

Number of grants (including subgrants) with the federal government:

      Current fiscal year (2014):_____;
      Fiscal year 2013:_____;
      Fiscal year 2012:_____.

Federal agencies with which federal grants are held:

      Current fiscal year (2014):_____;
      Fiscal year 2013:_____;
      Fiscal year 2012:_____.

List of subjects of federal grants(s) (for example, materials research, sociological study, software design, etc.):

      Current fiscal year (2014):_____;
      Fiscal year 2013:_____;
      Fiscal year 2012:_____.

Aggregate dollar value of federal grants held:

      Current fiscal year (2014):_____;
      Fiscal year 2013:_____;
      Fiscal year 2012:_____.

Bryan McGrath

Digitally signed by Bryan McGrath
DN: cn=Bryan McGrath, o=The FerryBridge Group LLC, ou=Managing Director, email=bmcgrath@ferrybridgegroup.com, c=US
Date: 2014.07.10 21:19:28 -04'00'

NOT FOR PUBLICATION UNTIL RELEASED BY
THE HOUSE ARMED SERVICES COMMITTEE
SEAPOWER AND PROJECTION FORCES
SUBCOMMITTEE

STATEMENT OF

VICE ADMIRAL PAUL A. GROSKLAGS
PRINCIPAL MILITARY DEPUTY, ASSISTANT SECRETARY OF THE NAVY
(RESEARCH, DEVELOPMENT AND ACQUISITION)

AND

MR. MARK D. ANDRESS
ASSISTANT DEPUTY CHIEF OF NAVAL OPERATIONS
(N2/N6)

AND

BRIG GEN JOSEPH T. GUASTELLA
DEPUTY DIRECTOR, REQUIREMENTS
(JOINT CHIEFS OF STAFF/J8)

BEFORE THE

SEAPOWER AND PROJECTION FORCES SUBCOMMITTEE

OF THE

HOUSE ARMED SERVICES COMMITTEE

ON

UNMANNED CARRIER LAUNCHED AIRBORNE SURVEILLANCE
AND STRIKE PROGRAM

JULY 16, 2014

NOT FOR PUBLICATION UNTIL RELEASED BY
THE HOUSE ARMED SERVICES COMMITTEE
SEAPOWER AND PROJECTION FORCES

INTRODUCTION

Mr. Chairman, Ranking Member McIntyre, and distinguished members of the Subcommittee, we thank you for the opportunity to appear before you today to discuss the Department of the Navy's (DoN) Unmanned Carrier Launched Airborne Surveillance and Strike (UCLASS) program. The UCLASS program will be an important addition to the Department of Defense's broad portfolio of programs that serve the near-term ISR needs of the nation and the joint warfighters. UCLASS, therefore, must be viewed as one of many assets that provide various capabilities including ISR, persistence flexible mission payloads (sensors and weapons) and enhanced survivability, to name a few.

The United States is a maritime nation with global responsibilities. Our Navy and Marine Corps' persistent presence and multi-mission capability represent U.S. power projection across the global commons. Navy and Marine Corps forces move at will across the world's oceans, seas and littorals, and they extend the effects of the sea-base deep inland. Naval Aviation provides our nation's leaders with "offshore options" where needed, when needed. We enable global reach and access, regardless of changing circumstances, and will continue to be the nation's preeminent option for employing deterrence through global presence, sea control, mission flexibility and when necessary, armed interdiction. The Navy and Marine Corps provide an agile strike and amphibious power projection force in readiness, and such agility requires that the aviation arm of our naval strike and expeditionary forces remain capable in the future threat environment. UCLASS will enhance our Naval and Joint Force capabilities by providing the carrier air wing with organic persistent Intelligence, Surveillance, Reconnaissance, and Targeting (ISR&T) and precision strike capability.

Unmanned Carrier Launched Airborne Surveillance and Strike (UCLASS) System

The UCLASS system is the next step in the Navy's evolutionary integration of unmanned air systems into the carrier strike group operational environment. It will provide a persistent, aircraft carrier-based, ISR&T and precision strike capability with inherent provisions for growth in mission capability, keeping UCLASS relevant long into the future.

The DoN is fully committed to UCLASS. Our Fiscal Year 2015 President's Budget requests $403.0 million in RDT&E,N for system development efforts to meet Joint Requirements Oversight Council (JROC) direction to expedite fielding of an early operational capability. The JROC has re-affirmed, as recently as February 2014, the need for rapid fielding of an affordable, adaptable, carrier-based ISR&T platform with future precision strike capability.

Warfighter representatives from the Chief of Naval Operations (CNO) staff, Commander of Naval Air Forces, and United States Fleet Forces Command collaborated over the last four years to ensure alignment of all UCLASS requirements. The CNO signed the Capabilities Development Document (CDD) in April 2013. Key Performance Parameters (KPPs) and Key System Attributes (KSAs) have remained stable and unchanged since that time.

UCLASS KPPs and KSAs address affordability, persistence, sensor payload, weapons payload (including future growth capability), survivability (including future growth capability), and aerial refueling (give and receive). UCLASS is required to be fully integrated within the current carrier air-wing and sustainable onboard an aircraft carrier. It will also have the ability to pass command and control information along with sensor data to other aircraft, naval vessels, and ground forces. Sensor data will be transmitted to exploitation nodes afloat and ashore. Interfaces will be provided with existing ship and land-based command and control systems, as well as processing, exploitation, and dissemination systems.

Based on the technology advancement and maturation demonstrated via the UCAS-D program, combined with insight gained through recent UCLASS Preliminary Design Reviews (PDRs) conducted with Northrop-Grumman, Lockheed-Martin, Boeing, and General Atomics, the Navy is confident that a government-industry team has the ability to deliver a UCLASS system that meets Service-approved CDD requirements within planned cost and schedule.

Significant reduction in FY15 UCLASS funding or a program pause for further review of UCLASS requirements will significantly delay source-selection activities, award of a development contract to industry, and will negatively impact delivery of an early operational capability. Any significant delay at this point in the program will also jeopardize continued investment and/or participation by one or more industry partners.

DoD, in concert with Congress, has spent the last four years in assessment of the UCLASS performance requirements, leading to the balanced capability reflected in the recently released draft Request for Proposal. In parallel, the Navy has developed an acquisition strategy that balances affordability and expediency with the ability to cost effectively expand UCLASS capabilities to address future threats.

United States Navy
Biography

Vice Admiral Paul A. Grosklags
Principal Military Deputy
Assistant Secretary of the Navy for Research, Development, and Acquisitions

Vice Adm. Grosklags is a native of DeKalb, Ill. After being designated a naval aviator in October 1983, he immediately reported to Training Squadron Three at North Whiting Field in Milton, Fla., as a T-34C flight instructor.

Grosklags served operational tours with Helicopter Antisubmarine Squadrons 34 and 42, where he flew the SH-2F and SH-60B, respectively. Grosklags made multiple deployments with the USS John Hancock (DD 981), USS Donald B. Beary (FF 1085), USS Comte de Grasse (DD 974), and USS Leyte Gulf (CG 55). He later served as both executive and commanding officer of Helicopter Training Squadron Eighteen 18.

Grosklags' acquisition tours include engineering test pilot and assignments as MH-60R assistant program manager for systems engineering, H-60 assistant program manager for test and evaluation, MH-60R deputy program manager, and ultimately as program manager for Multi-Mission Helicopters (PMA-299), during which time the MH-60R was successfully introduced to the fleet. Grosklags also served as operations officer and subsequently as deputy Program Executive Officer for Air Anti-Submarine Warfare, Assault and Special Mission Programs (PEO(A)).

Grosklags has served flag tours as commander, Fleet Readiness Centers and Naval Air Systems Command (NAVAIR) assistant commander for Logistics and Industrial Operation, NAVAIR vice commander, and PEO(A). In July 2013, he assumed responsibilities as principal military deputy for the Assistant Secretary of the Navy (Research, Development & Acquisition).

Grosklags graduated from the U.S. Naval Academy in 1982, is a graduate of the U.S. Naval Test Pilot School Class 99, and holds a Master of Science degree in Aeronautical Engineering from the Naval Postgraduate School. He has more than 5,000 military flight hours in numerous types of rotary and fixed-wing aircraft. Grosklags is a proud but humble co-owner of the Green Bay Packers and works weekends providing free labor on his wife's fish farm.

Mark D. Andress

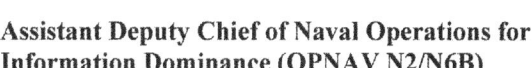

Assistant Deputy Chief of Naval Operations for
Information Dominance (OPNAV N2/N6B)

Mr. Andress serves as the Assistant Deputy Chief of Naval Operations for Information Dominance (OPNAV N2/N6B). The mission of N2/N6 is to manage requirements, policy and resources for the Navy's Communications, Networks, ISR, Oceanography, Space, Cyber, Electronic Warfare and Command and Control capabilities.

Mr. Andress held two previous assignments at the Pentagon beginning in 2009 as a Defense Intelligence SES. He oversaw N2/N6's planning, programming, budgeting and execution for annual investments of over $16 billion, more than 180 programs and 32,000 military, civilian and contractor personnel. Prior to that assignment, he was the division director for the Navy's Command, Control, Intelligence, Surveillance and Reconnaissance (C2ISR) programs. Major programs included the E-2 and E-6 aircraft, airborne tactical networks and command and control/intelligence systems.

From 2004 to 2009, Mr. Andress served in a variety of leadership positions at the Office of Naval Intelligence. He was the Executive Director of the Hopper Information Services Center, a command of over 600 personnel delivering information technology and ISR technical support to the Navy. He also served as Deputy Director of ONI's Maritime Intelligence Operations Center with over 500 personnel from across the Intelligence Community delivering direct support to Navy operations. From 2000 to 2004, Mr. Andress was a program manager at Military Sealift Command (MSC) overseeing the their shipboard information systems programs.

A former U.S. Navy Surface Warfare Officer and U.S. Coast Guard licensed merchant mariner, he spent the first 12 years of his career at sea either in uniform or in the commercial maritime industry.

Mr. Andress completed his bachelor's of science degree in aerospace engineering from Texas A&M University in 1988. He completed his master's of engineering management degree from George Washington University in 2002.

BIOGRAPHY

UNITED STATES AIR FORCE

BRIGADIER GENERAL JOSEPH T. GUASTELLA

Brigadier General Joseph T. Guastella is the Deputy Director for Requirements (J8), Joint Staff, the Pentagon, Washington, D.C. He is responsible for assisting the Chairman of the Joint Chiefs of Staff in identifying, assessing and approving joint military requirements to meet the national military strategy; and in ensuring the consideration of trade-offs among cost, schedule, and performance objectives for those requirements. In this role, he orchestrates Joint Staff support of the capabilities development process through the Joint Capabilities Integration and Development System. Additionally, he oversees the validation for joint urgent operational needs leading to the fielding of rapid solutions for combatant commanders.

General Guastella entered the Air Force in 1987 as a graduate of the U.S. Air Force Academy. He's flown the F-16 Fighting Falcon and A-10 Thunderbolt II, served multiple combat tours, and instructed at the U.S. Air Force Fighter Weapons School. He commanded the 555th Fighter Squadron, 'Triple Nickel', Aviano Air Base, Italy and the 20th Fighter Wing at Shaw Air Force Base, S.C.

Prior to his current assignment, he was the Commander, 455th Air Expeditionary Wing, Bagram Airfield, Afghanistan.

EDUCATION

1987 Bachelor of Science in Astronautical Engineering, U.S. Air Force Academy, Colorado Springs, Colo.
1994 Squadron Officer School, Maxwell AFB, Ala.
1997 Master of Science in Aero Science Technology, Embry Riddle University
2001 Air Command and Staff College, Maxwell AFB, Ala.
2006 Master of Science in National Security Strategy, National War College, Fort McNair, Washington D.C.
2011 Senior Executive Fellows program, John F. Kennedy School of Government, Harvard University, Cambridge, Mass.

ASSIGNMENTS
1. August 1987 - January 1989, student pilot, 80th Flying Training Wing, Sheppard AFB, Texas
2. January 1989 - September 1989, F-16 initial training, Luke AFB, Ariz.
3. September 1989 - May 1992, squadron electronic combat pilot and assistant weapons officer, 526th Tactical Fighter Squadron, Ramstein Air Base, Germany
4. May 1992 - May 1993, instructor pilot, assistant flight commander and assistant weapons officer, 35th Fighter Squadron, Kunsan Air Base, Korea

5. May 1993 - April 1994, standardization and evaluation flight examiner, 526th TFS, Ramstein AB, Germany

6. April 1994 - January 1995, standardization and evaluation flight examiner, 555th FS, Aviano AB, Italy

7. Jan 1995 - June 1995, student, fighter Weapons School, Nellis AFB, Nev.

8. June 1995 - December 1996, squadron weapons and tactics officer and assistant wing weapons officer, 555th FS, Aviano AB, Italy

9. December 1996 - August 2000, Instructor, F-16 Fighter Weapons Instructor Course, Nellis AFB, Nev.

10. September 2000 - June 2001, student, Air Command and Staff College, Maxwell AFB, Ala.

11. June 2001 - August 2002, action officer, Joint Strike Fighter and Combat Identification Programs, Headquarters Air Force, Directorate of Operational Requirements, Washington, D.C.

12. August 2002 - October 2003, operations officer, 555th FS, Aviano AB, Italy

13. October 2003 - July 2005, commander, 555th FS, Aviano AB, Italy

14. August 2005 - July 2006, student, National War College, Fort Lesley J. McNair, Washington, D.C.

15. June 2006 - June 2007, U.S. CENTAF, A3 forward and CAOC Director of Operations, Southwest Asia

16. June 2007 - August 2008, Deputy Director CAPSTONE, National Defense University, Fort Lesley J. McNair, Washington D.C.

17. August 2008 - October 2008, vice commander, 20th FW, Shaw AFB, S.C.

18. October 2008 – June 2010, commander, 20th FW, Shaw AFB, S.C.

19. June 2010 – July 2011, Chief Program Integration Division, Directorate of Programs, DCS Strategic Plans and Programs, Pentagon Washington D.C.

20. July 2011 – June 2012, Deputy Director of Programs, Office of the Deputy Chief of Staff for Strategic Plans and Programs, Headquarters U.S. Air Force, Washington, D.C.

21. July 2012 – July 2013, commander, 455 AEW, Bagram, Afghanistan

22. July 2013 – present, Deputy Director of Requirements (J8), Joint Staff, Washington D.C.

SUMMARY OF JOINT ASSIGNMENTS

1. June 2006 - June 2007, U.S. Air Forces Central Command, A3 forward and Combined Air Operations Center Director of Operations, Southwest Asia, as a colonel

2. June 2007 - August 2008, Deputy Director CAPSTONE, National Defense University, Fort Lesley J. McNair, Washington D.C., as a colonel

3. July 2012 - July 2013, Commander 455 Air Expeditionary Wing, Bagram, Afghanistan, Regional Command East, as brigadier general

4. July - Present, Deputy Director for Requirements, (J8) Joint Staff as brigadier general

FLIGHT INFORMATION
Rating: Command Pilot
Flight Hours: More than 4,000 (1000+ combat)
Aircraft Employed: F-16C/D, A-10C

MAJOR AWARDS AND DECORATIONS
Defense Superior Service Medal
Legion of Merit with one oak leaf cluster
Bronze Star Medal with one oak leaf cluster
Meritorious Service Medal with one oak leaf cluster
Air Medal with eleven oak leaf clusters
Aerial Achievement Medal
Air Force Commendation Medal with three oak leaf clusters
Air Force Achievement Medal
Joint Meritorious Unit Award
Air Force Outstanding Unit Award with four oak leaf clusters
Air Force Organizational Excellence Award
Combat Readiness Medal with two oak leaf clusters

(Current as of August 2013)

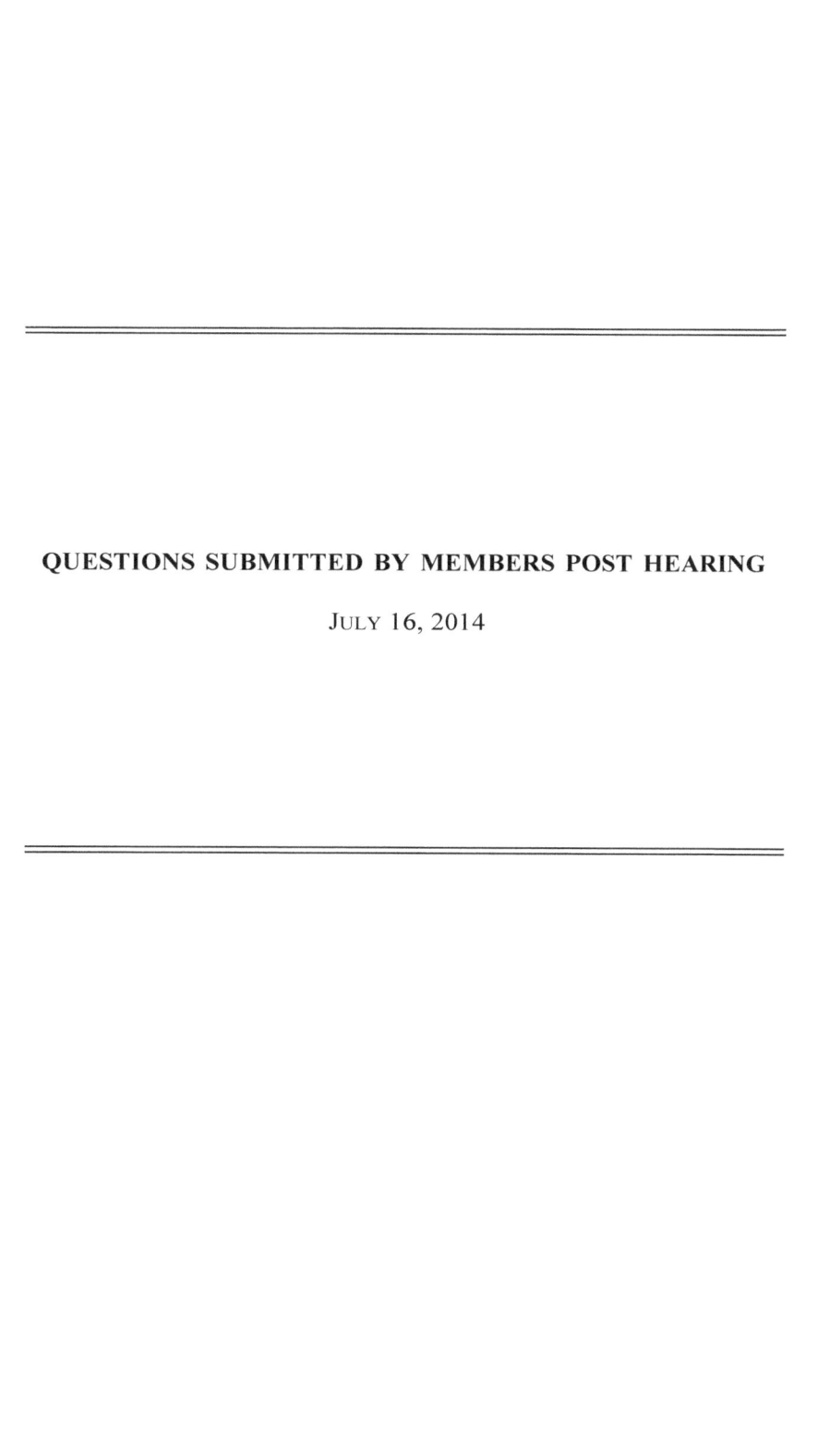

QUESTIONS SUBMITTED BY MEMBERS POST HEARING

JULY 16, 2014

QUESTIONS SUBMITTED BY MR. HUNTER

Mr. HUNTER. How does this platform fit into current or anticipated operational plans? Specifically, which mission is it most important that this platform be capable of performing? Are we seeking UCLASS to be a naval force multiplier or focused for deep strike?

Admiral GROSKLAGS. Persistent sea-based Intelligence, Surveillance, Reconnaissance, and Targeting (ISR&T) with precision strike is the most critical gap that UCLASS will fill. This was reinforced in a combined USAFRICOM, USCENTCOM and USSOCOM Joint Emergent Operational Need (JEON), in which UCLASS was deemed a suitable solution. UCLASS will fully integrate with the Carrier Air Wing to support a myriad of missions from permissive counter-terrorism operations to missions in low-end contested environments, to providing enabling capabilities for high-end denied operations, as well as supporting organic naval missions. The UCLASS initial Concept of Operations (CONOPS) includes long dwell surveillance and targeting for extended range weapons, and precision strike against time sensitive targets.

Mr. HUNTER. What sort of ordnance do you see a strike oriented solution carrying? What types of targets do you envision it striking? Are these missions already within the F/A–18 or F–35C capability? How would a strike oriented UCLASS compliment or overlap with capabilities offered by those platforms?

Admiral GROSKLAGS. Given the projected speed of UCLASS and most probable target sets, the weapons most valued by fleet requirements are the 500-pound class Laser Joint Direct Attack Munitions (JDAM) and the upgraded Small Diameter Bomb (SDB II). The internal weapons capacity requirement of greater than 1000 pounds of ordnance at threshold was determined to address specific classified scenarios and compliments the Carrier Air Wing capability. An objective growth criteria for 2,000 pounds of LJDAMs and/or SDBIIs is also specified. Additionally, the UCLASS Air Vehicle will have a minimum of two external hard-points each provisioned to carry 3,000 pounds, suitable for most weapons currently employed by the Carrier Air Wing.

Mr. HUNTER. Do UCLASS budget estimates provided to Congress this year envision a low-cost, affordable acquisition strategy or do they envision development of a presumably more expensive, highly stealthy platform?

Admiral GROSKLAGS. The UCLASS budget profile, as outlined in the President's Budget for FY 2015, provides adequate funding to achieve the Early Operational Capability with growth capabilities outlined in the current acquisition strategy. Affordability is a key component of the UCLASS acquisition strategy.

Mr. HUNTER. What is your specific endurance requirement for UCLASS? Can you have it all—long endurance and high stealth—or is there a tradeoff? What considerations need to be weighed between a 14 hour endurance solution with little or no strike capability and a 10 hour endurance solution with significant strike capability?

Admiral GROSKLAGS. The endurance requirement, as part of the Persistence Key Performance Parameter, is to provide two unrefueled orbits (24/7 coverage) at a radius of 600 nm from the aircraft carrier, or one unrefueled orbit at 1200 nm per UCLASS system, or for each air vehicle to fly an unrefueled maximum range profile (out and back) to 2000 nm to perform strike missions.

The designs of aircraft which can meet the UCLASS requirements are driven by unrefueled endurance/range; senor payload weight/volume; weapons payload; low-observable characteristics; in-flight refueling provisions for both fuel give and receive; and, as important as any of the preceding, the constraints of operating from an aircraft carrier including consideration of such things as structural loads for launch/recovery, landing area (width), and deck spotting factor. Even with all of the above design constraints, through two years of close engagement with industry, the DoN is very confident that affordable UCLASS aircraft with 14 hours of unrefueled endurance and a high degree of low observability are possible and will be proposed by industry in response to our forthcoming request for proposals.

A 14 hour endurance UCLASS will be able to carry 1000 to 2000 pounds of internal weapons. One thousand pounds is the minimum requirement and the potential to attain 2000 pounds at EOC will be determined by specific vender proposals. This payload is sufficient to meet mission requirements as defined in the UCLASS Design Reference Missions. In addition the UCLASS will have a minimum of two 3000 pound external hard-points capable of handling the majority of weapons in the current CVW inventory.

———

QUESTIONS SUBMITTED BY MR. LANGEVIN

Mr. LANGEVIN. Looking at the vast array of ISR assets available to a carrier strike group commander—space, airborne both from land and sea, and undersea—what would be the "secret sauce" that an ISR-centric, 14-hour UCLASS would bring that an 8–10 hour UCLASS could not?

Mr. O'ROURKE. In preparation for the hearing, I received briefings on the UCLASS program from the Navy and from each of the four firms that are currently competing for the program. The Navy stated that its analysis of alternatives (AOA) for the UCLASS program concluded that an 8-hour UCLASS would have effectiveness comparable to a 14-hour UCLASS in three of the four tactical situations that were examined in the AOA, and somewhat less effectiveness than a 14-hour UCLASS in one of the four tactical situations. The Navy stated that the AOA concluded that an 8-hour UCLASS would have a higher life-cycle cost than a 14-hour UCLASS. Life-cycle cost, the Navy stated, included development, production, and operation and support (O&S) costs. Based on the briefings I received from industry, my sense is that some in industry might agree (or at least not disagree) with these findings, while others might disagree or argue that the AOA did not examine the right set of tactical situations. In my prepared statement for the hearing, I stated that:

The specific tactical situations that were examined in the UCLASS AOA are related to the program's current operational requirements. Assessing alternative operational requirements for the UCLASS program could involve examining potential outcomes in other tactical situations that may not have been considered in the AOA. A broader analysis might examine how changes in UCLASS operational requirements might affect estimated outcomes in campaign-level, force-on-force situations, rather than in specific tactical situations. (Statement of Ronald O'Rourke, Specialist in Naval Affairs, Congressional Research Service, Before the House Armed Services Committee, Subcommittee on Seapower and Projection Forces, on Unmanned Carrier-Launched Airborne Surveillance and Strike (UCLASS) Requirements Assessment, July 16, 2014, pp. 5–6.)

Mr. LANGEVIN. Looking at the vast array of ISR assets available to a carrier strike group commander—space, airborne both from land and sea, and undersea—what would be the "secret sauce" that an ISR-centric, 14-hour UCLASS would bring that an 8–10 hour UCLASS could not?

Mr. MARTINAGE. The short answer is that there is no "secret sauce" that a 14-hour UCLASS would bring that an 8–10 hour UCLASS with significantly enhanced survivability and strike capacity could not. The Navy's rationale for 14 hours of unrefueled endurance centers on its stated requirement for 24-hour intelligence, surveillance and reconnaissance (ISR) support of—or "maritime domain awareness" around—the carrier strike group (CSG). Putting aside survivability and strike capacity issues and looking narrowly at just the MDA mission (which, I would argue, is the central problem with UCLASS requirements), the only "advantage" conferred by 14 hours of unrefueled endurance is the ability to bridge the closed or "overnight" period of the canonical 12 hour "deck day" without land-based airborne tanker support.

If the deck day were extended by three to four hours, or if aircraft were arranged on the flight deck at the close of normal air operations to support one or two overnight recoveries, then this so-called advantage vanishes. Of course, if land-based aerial refueling is available, which is axiomatically true during joint combat operations, the 8–10 hour air vehicle could be "tanked" during the night rather than recover to the carrier. It is essential to note that if the carrier is supporting power projection operations in anti-access environments anticipated in the 2020s, which is when UCLASS would field, it would likely be standing off some 1,000–1,500 nautical miles (or more) from an adversary's coast. In which case, the only practical way to sustain persistent ISR-strike operations would be to refuel UCLASS inflight from Air Force tankers operating several hundred miles closer to the battlespace, just outside the range of adversary air interceptors and surface-to-air missiles. A non-refuelable UCLASS, even one with 14 hours unrefueled endurance, would offer marginal utility in such scenarios.

Equally important to note is that the Navy already has two other UAS programs of record underway that will yield overlapping, multi-tiered (strategic- and tactical-

level) persistent MDA in support of CSG operations. Indeed, the MQ–4C Triton, a
marinized version of the Air Force's RQ–4 Global Hawk strategic surveillance UAS
with over 30 hours unrefueled endurance, was designed expressly for the "broad
area maritime surveillance" mission (hence, its previous "BAMS" moniker). And
while the Navy is currently planning to acquire some 68 Tritons, it would take just
three of these aircraft (two operational, one spare) to sustain 24-hour MDA in support of a CSG. Triton is complemented in the MDA domain by the MQ–8C Fire
Scout, an unmanned helicopter with over 12 hours unrefueled endurance capable of
operating from any air-capable surface combatant (e.g., carriers, destroyers, cruisers, littoral combat ships). Together, Triton and Fire Scout will arguably generate
a surfeit of MDA capacity over the near-term. Thus, it makes little sense for the
MDA mission—much less the readily dispelled notion of an MDA "capability gap"—
to drive UCLASS requirements.

The more important issue this question raises is the opportunity cost of that 14-
hour endurance in terms of reduced payload capacity and flexibility, survivability,
and growth potential. The cost of 14 hours of unrefueled endurance is most likely
about 1,000–3,000 lbs of weapons (or other mission payloads), the inability to carry
stand-off weapons both currently in the inventory and in development, and a significant increase in presented radar cross section at relevant threat frequencies.

Mr. LANGEVIN. Looking at the vast array of ISR assets available to a carrier
strike group commander—space, airborne both from land and sea, and undersea—
what would be the "secret sauce" that an ISR-centric, 14-hour UCLASS would bring
that an 8–10 hour UCLASS could not?

Mr. BRIMLEY. The short answer is, no secret sauce would be provided from an
ISR-centric, 14-hour UCLASS. As discussed during the hearing, the types of permanent design decisions that would be required to field a 14-hour UCLASS would preclude the kind of payload and limited stealth that would be required to provide a
meaningful increase to the striking power of the carrier air wing. As I described
during the hearing, I am not a former naval officer and I am not an aircraft design
expert—when I engage in a requirements discussion my perspective is that of a civilian defense strategist. That is to say, I concern myself with several threshold
questions, including:

1. Will the platform provide a future Commander-in-Chief better military options
during a crisis? 2. Will it help address pressing gaps in U.S. defense strategy and
planning? 3. Does it enable forward U.S. forces to present a stronger conventional
deterrent and, if necessary, help ensure U.S. forces can defeat a plausible adversary? 4. Will the program help underwrite the confidence of our allies and partners?
5. Does it reflect measured judgments regarding mid- to long-term requirements for
U.S. defense? 6. And finally, does the program help ensure America's military-technical dominance in an increasingly competitive environment?

I don't think a very limited system designed to provide ISR to the carrier mean-
ingfully addresses any of the above questions. Far and away the most pressing challenge facing the U.S. Navy is finding ways to project and sustain combat power in
the face of adversary ballistic and cruise missile technology that could hold at risk
our aircraft carriers well beyond the unrefueled range of their strike aircraft. The
original requirements for an unmanned combat aerial vehicle (UCAV) date back to
the 2006 Quadrennial Defense Review. I believe the original conception of harnessing the unmanned revolution to provide an asymmetric and disruptive capability that would ensure the combat relevance of the carrier air wing to a plausible
high-end challenge were correct. We need a system that has broadband, all-aspect
stealth, is capable of automated aerial refueling, with an integrated surveillance
and strike capability. My argument is that these original requirements were largely
correct, and the recent deviations from this to a more limited ISR role reportedly
described in the draft UCLASS RFP are not wise given the projected security environment.

Finally, as we talked about during the hearing, it is not as if several of my colleagues and I were arguing for a capability that is well outside the realm of the
possible. We all observed the multiple recent successful tests of the Navy's X–47B,
a stealthy flying wing design that will likely succeed in air-to-air refueling tests as
well. I am confident that we can build and field the kind of system that the Navy
will need to give a future Commander-in-Chief real deterrent and carrier-based
strike options that I believe he or she will need in the years ahead.

Mr. LANGEVIN. Looking at the vast array of ISR assets available to a carrier
strike group commander—space, airborne both from land and sea, and undersea—
what would be the "secret sauce" that an ISR-centric, 14-hour UCLASS would bring
that an 8–10 hour UCLASS could not?

Mr. MCGRATH. Thank you, Representative Langevin, for your continuing interest
in this matter and for this excellent question.

You correctly list a number of ISR sensors to which a Strike Group Commander has access. The array is considerable, and in some respects, overlapping. To the extent that there is any "secret sauce" available, it all revolves around the question of "who controls the asset?" If the CSG Commander had a 14 hour ISR privileged UCLASS at his disposal and UNDER HIS COMMAND, it gives him additional operational flexibility versus assets "owned and operated" by some other commander. Under current Command and Control (C2) schemes, the CSG Commander must request assets from others to fill this mission, and this is something no commander enjoys. Having instant and untrammeled control of a capability is always better than "access" to someone else's asset.

This said, the CSG Commander relies on others to provide him with a lot of capabilities. He does not own the P–8's or MQ–4's that provide him with support. He does not own satellite assets that give him support. He often relies on inorganic tanking from USAF refueling assets. The point is, the aircraft carrier deck is already a crowded place, so taking up valuable real estate simply to provide the CSG Commander with an asset solely under his control—the opportunity cost of which is moving more slowly and ineffectively to combat capability in a contested environment—seems imprudent when there are a plethora of other ways to get the desired information to the Strike Group.

Mr. LANGEVIN. Why is air-to-air refueling not a threshold requirement? What effect would having the ability to refuel have on the requirement—for instance, would the endurance requirement come down in a CONOPS where a UCLASS platform was able to refuel after takeoff, like current manned missions?

Admiral GROSKLAGS. The UCLASS Air Vehicle will be required to be provisioned for aerial refueling at threshold. All vendors are required to meet the requirements set forth in the Persistence KPP which call for two 600 nm 24/7 orbits or one 1200 nm 24/7 orbit or one 2000nm strike; all of which must be conducted unrefueled. Due to the compressed test and evaluation period and requirements outlined in the affordability Key Performance Parameter, aerial refueling, which includes both giving and receiving fuel, would not be achievable by the 2020 Early Operational Capability (EOC) deployment and will be implemented in the future based on Fleet requirements/demand. Complete implementation of aerial refueling at threshold would not affect current threshold endurance requirements. Additionally, including air-to-air refueling at threshold would add technical difficulty and cost to the development process making the program unaffordable. The tanker fleet would also require additional development and test, adding cost and time to the program.

Mr. LANGEVIN. I'm sure each of you are aware of the public reports of several nations developing advanced radar systems and radar networks specifically designed to defeat low-observable platforms. Given the pace of development and the proliferation of air defense radar systems in the past, how confident are you that the levels of low-observability across key frequencies that the Navy is planning to require would be sufficient for UCLASS to conduct the full range of envisioned missions through the life of the platform?

Admiral GROSKLAGS. When developing the Air Vehicle survivability specifications, a broad range of current and future threat systems were evaluated. This assessment looked at a full range of scenarios including shore based and maritime A2AD threats. The Early Operational Requirement threshold capabilities, future growth, requirements, and objective criteria were based on this assessment and the Concept of Operations which utilizes UCLASS as part of a fully integrated Carrier Air Wing/Carrier Strike Group.

Mr. LANGEVIN. Can you give us an example of a mission that a 14-hour endurance UCLASS could accomplish that a 8–10 hour vehicle or other assets, whether sea, air, or space, could not?

General GUASTELLA. Persistence is a key attribute for the Unmanned Carrier Launched Airborne Surveillance and Strike (UCLASS) system. Based on extensive endurance and aerial refueling analyses, a 14-hour vehicle is the most cost effective approach to meet operational requirements. UCLASS must integrate into the standard carrier (CVN) flight cycle, and while an air vehicle with 8–10 hour endurance may have the same surveillance and strike capabilities, the shorter endurance will require costly aerial refueling to integrate into CVN operations. Avoiding the reliance on aerial refueling provides a cost advantage and reduced operational risk. A 14-hour unrefueled endurance allows 24-hour coverage within a standard fly day and the greater persistence and range translate to greater operational flexibility for the Carrier Strike Group and operational commanders.

Mr. LANGEVIN. Why was 1,000 pounds chosen as a threshold strike capability? What does that translate to as far as weapons capabilities, including standoff weapons? How does the weight relate to volume requirements? And what requirement is this strike capability designed to address?

115

General GUASTELLA. The Unmanned Carrier Launched Airborne Surveillance and Strike (UCLASS) Analysis of Alternatives (AOA) examined several performance recommendations that drove the development of the Key Performance Parameters (KPPs) outlined in the draft Capabilities Development Document (CDD). The internal weapons capacity of > 1000 pounds at threshold was determined by several factors including endurance, survivability, carrier integration and, most importantly, targets serviced. Based on this analysis, UCLASS' precision strike capability will be designed to address specific classified scenarios for both today's missions and future threats. In these scenarios, the 500-pound class Laser Joint Direct Attack Munitions (LJDAM) and the upgraded Small Diameter Bomb (SDB II) are the weapons of choice. Each of these weapons enhance ''maneuverability'' when employed against a moving target like a ship or vehicle, and SDBII provides a significant standoff capability. The threshold requirement of 1000 lbs internal, with objective growth approaching 2000 lbs allows for 2–4 LJDAM or 4–8 SDBIIs. This number of weapons provides sufficient precision strike capability to service the target sets outlined in the AOA. Additionally, the UCLASS air vehicle is planned to have at least two external hard-points, provisioned for additional 3000 lbs of carriage each, to carry a majority of the weapons currently employed by the Carrier Air Wing.

Mr. LANGEVIN. I'm sure each of you are aware of the public reports of several nations developing advanced radar systems and radar networks specifically designed to defeat low-observable platforms. Given the pace of development and the proliferation of air defense radar systems in the past, how confident are you that the levels of low-observability across key frequencies that the Navy is planning to require would be sufficient for UCLASS to conduct the full range of envisioned missions through the life of the platform?

General GUASTELLA. We are confident that the Joint Requirements Oversight Council (JROC) survivability requirements for the Unmanned Carrier Launched Airborne Surveillance and Strike (UCLASS) program are sufficient. The Joint unmanned aerial system portfolio includes systems with various levels of performance, survivability, basing options and missions. The UCLASS will play a key role in providing carrier-based, persistent intelligence, surveillance, reconnaissance, and precision strike capability within this portfolio.

———

QUESTIONS SUBMITTED BY MR. LARSEN

Mr. LARSEN. To what degree have you reviewed the Navy's UCLASS draft RFP and classified addenda? Are there aspects of the Navy defined survivability requirement or other requirements you find insufficient and why?

Mr. MARTINAGE. I reviewed earlier drafts of these materials while serving in the Department of the Navy. Questions about survivability cannot be adequately addressed at the unclassified level.

Mr. LARSEN. Since the Navy and OSD/Joint Staff vetted the UCLASS requirements, has additional information come to light to warrant a change to those requirements at this stage of the acquisition process?

Mr. MARTINAGE. Countering emerging anti-access/area-denial (A2/AD) challenges was OSD's original motivation for both starting Navy UCAS/UCAS–D in the 2006 Quadrennial Defense Review (QDR), and for providing the additional $2 billion in the FY11 Program and Budget Review. The need for a longer range, survivable, carrier-based air vehicle for ISR and strike in contested airspace was articulated throughout the 2010 QDR and affirmed in testimony to Congress on several occasions by senior Navy official in 2010 and 2011.

The current key performance parameters (KPPs) emerged from a highly contentious—and still unsettled—debate with DOD over the past two years. Among the competing schools of thought are those who seek a lower-end, carrier-based UAS optimized for counter-terrorism missions as a hedge against the potential loss of land-bases for armed UAVs such as the MQ–1 Predator and MQ–9 Reaper; those that believe a capability-gap exists with respect to maritime domain awareness (MDA) around the carrier strike group (CSG); those that are willing to dilute UCLASS requirements to reduce bureaucratic and cultural resistance within the naval aviation community to ''get something'' onto the carrier deck; and those, like myself, that fervently believe that a stealthy, air-refuelable ISR-strike UAS is needed to maintain the operational relevance of the carrier air wing in the face of emerging A2/AD threats. The latter would offer ''pan-conflict spectrum utility,'' meaning that it would be equally capable of the counter-terrorism, MDA, and counter-A2/AD power projection missions.

What has changed since this debate was first joined is a growing awareness within DOD and the national security community of the probable scale, scope, and pace of the unfolding A2/AD challenge. Meanwhile, the feared loss of land bases to support counter-terrorism operations in the Middle East, Central Asia, and Africa has not materialized and there are no signs that it will. While it is difficult to say whether counter-terrorism will be as prominent in the mid-2020s when UCLASS is scheduled to field as today, current trends suggest that the U.S. military will retain a wide range of options for basing long-range UAVs such as the extended-range MQ–9 Reaper and RQ–4 Global Hawk. In contrast, threats to the aircraft carrier and its embarked aircraft are clearly intensifying.

Although several countries around the world are fielding A2/AD capabilities, the pacing threat is China. In its most recent *Annual Report to Congress on Military and Security Developments involving the People's Republic of China,* DOD highlights myriad threats to the aircraft carrier including air-, sea-, and submarine-launched anti-ship cruise missiles; wake-homing torpedoes from a growing and increasingly capable submarine fleet; and long-range, anti-ship ballistic missiles (ASBMs). It states:

> *China is fielding a limited but growing number of conventionally armed, medium-range ballistic missiles, including the DF–21D anti-ship ballistic missile (ASBM). The DF–21D is based on a variant of the DF–21 (CSS–5) medium-range ballistic missile (MRBM) and gives the PLA the capability to attack large ships, including aircraft carriers, in the western Pacific Ocean. The DF–21D has a range exceeding 1,500 km and is armed with a maneuverable warhead.* (pp. 5–6.)

In addition to the carrier being potentially pushed well outside the unrefueled combat radius of its embarked fighters, the F–18E/F and F–35C will also confront increasingly deadly land- and sea-based integrated air defenses (IADS). Not only are modern IADS diffusing widely around the globe, they are also growing more lethal owing to several synergistic trends: more sensitive radars operating over wider frequency bands, increased resistance to electronic attack (e.g., jamming and spoofing), increased interceptor range, more advanced signal processing, and high-speed networking. Variants of the Russian-made S–300 (SA–10/20), for example, are already in service in about a dozen countries, including Algeria, Armenia, Azerbaijan, Belarus, Bulgaria, China, Slovakia, and Venezuela. Both Iran and Syria have repeatedly attempted to procure the S–300 from Russia. China has already fielded a dense, networked IADS. As the most recent *Annual Report to Congress on Military and Security Developments involving the People's Republic of China* states:

> *China's ground-based air defense A2/AD capabilities will likely be focused on countering long-range airborne strike platforms with increasing numbers of advanced, long-range SAMs. China's current air and air defense A2/AD components include a combination of advanced long-range SAMs—its indigenous HQ–9 and Russian SA–10 and SA–20 PMU1/PMU2, which have the advertised capability to protect against both aircraft and low-flying cruise missiles. China continues to pursue the acquisition of the Russian extremely long-range S–400 SAM system (400 km), and is also expected to continue research and development to extend the range of the domestic HQ–9 SAM to beyond 200km.* (p. 35)

Prospective adversaries are also investing in more capable air superiority fighters, outfitted with modern sensor systems and armed with beyond-visual-range (BVR) air-to-air missiles. These aircraft can be vectored—in some cases, in large numbers—to intercept U.S. aircraft based on rough targeting tracks developed by ground-based early warning radars. U.S. tanker aircraft will need to honor both the unrefueled radius of adversary fighters, and also the range of their BVR missiles, when establishing aerial refueling "tracks" (rendezvous points) for penetrating U.S. aircraft. Against a nation such as China, which has a growing force of air interceptors with unrefueled radii between 600–900 nautical miles, this would require U.S. tankers to stand off as much as 750–1,000 nautical miles. It is critical to note that this standoff distance exceeds the unrefueled radii of the F/A–18E/F, F–22 and F–35A/B/C; and thus, would effectively preclude a penetrating offensive role for the entire U.S. fighter force. No fact more vividly underscores the need to shift emphasis within the attack capability area from short-range, manned fighter aircraft to penetrating, long-range, manned and unmanned ISR-strike systems.

Responding to this growing appreciation of the intensifying A2/AD threat around the world, the Quadrennial Defense Review (QDR) stressed the need to improve U.S. power projection capability in contested environments. Consider the following excerpts from the QDR:

> *In the coming years, countries such as China will continue seeking to counter U.S. strengths using anti-access and area-denial (A2/AD) approaches and by employing other new cyber and space control technologies. Additionally, these and other states continue to develop sophisticated integrated air defenses that can restrict access and freedom of maneuver in waters and airspace beyond terri-*

*torial limits. Growing numbers of accurate conventional ballistic and cruise mis-
sile threats represent an additional, cost-imposing challenge to U.S. and partner
naval forces and land installations.* (pp. 6–7)

   *As the Department rebalances toward greater emphasis on full-spectrum oper-
ations, maintaining superior power projection capabilities will continue to be
central to the credibility of our Nation's overall security strategy.* (p. 19)

   *The Department's investments in combat aircraft, including fighters and long-
range strike, survivable persistent surveillance, resilient architectures, and undersea warfare will increase the Joint Force's ability to counter A2/AD challenges.* (p. 36)

This recognition of the need to adapt U.S. power projection capabilities to address
current and emerging A2/AD challenges was re-affirmed recently in the independent
review of the QDR conducted by National Defense Panel created by Congress. That
panel, chaired by William Perry and John Abizaid, unanimously concluded that:

   *We believe it is also critical to ensure that U.S. maritime power projection ca-
pabilities are buttressed by acquiring longer-range strike capability—again,
manned or unmanned (but preferably stealthy)—that can operate from U.S. aircraft carriers or other appropriate mobile maritime platforms to ensure precise,
controllable, and lethal strike with greater survivability against increasingly
long-range and precise anti-ship cruise and ballistic missiles. (p. 43.)*

To conclude, the current UCLASS requirements as endorsed by the Joint Requirements Oversight Council (JROC) appear increasingly misaligned with DOD's own
threat assessment and articulation of the Nation's overall security strategy in the
QDR (as well as in the Defense Strategic Guidance previously approved by the
President).

   As I stated in my testimony, an assessment of UCLASS requirements should
begin with a very simple question: what is the core operational challenge facing carrier-based power projection? Although the Navy and the joint force more broadly
have multiple means of providing MDA around the carrier strike group and to identify targets for attack by relatively short-range, manned fighters in low-to-medium
threat environments, that is the focus of the JROC-approved KPPs. In my view, the
far more pressing challenge will be projecting power effectively when the carrier is
compelled to standoff at considerable distance (e.g., 1,000–1500 nm) from an adversary's coast, and then find and engage targets defended by modern IADS. UCLASS
requirements should be adjusted now to address that extant and intensifying challenge.

   Mr. LARSEN. To what degree have you reviewed the Navy's UCLASS draft RFP
and classified addenda? Are there aspects of the Navy defined survivability requirement or other requirements you find insufficient and why?

   Mr. BRIMLEY. I have not reviewed the UCLASS draft RFP or addenda, both of
which are classified. I base my opinions on the open-source reporting that I have
confidence in, including official statements from Navy officials—all of which describe
requirements for an ISR-centric platform. As we discussed during the hearing, I believe an ISR-centric capability will do little or nothing to address the main challenge
facing the carrier air wing and the carrier strike group more broadly, which is the
need to provide persistent combat strike power over long ranges in the face of adversary systems designed to target our aircraft carriers well outside the unrefueled
radii of the air wing. Addressing that capability gap, rather than add a redundant
ISR capability, strikes me as a more prudent way to invest limited taxpayer resources.

   Mr. LARSEN. Since the Navy and OSD/Joint Staff vetted the UCLASS requirements, has additional information come to light to warrant a change to those requirements at this stage of the acquisition process?

   Mr. BRIMLEY. The requirement for a stealthy, refuelable unmanned carrier-based
strike aircraft was relatively constant since the 2006 Quadrennial Defense Review,
which directed the Navy to "develop an unmanned longer-range carrier-based aircraft capable of being air-refueled to provide greater standoff capability, to expand
payload and launch options, and to increase naval reach and persistence." Since
then, both the 2010 QDR, the 2012 Defense Strategic Guidance, and the 2014 QDR
all restated the need to develop capabilities that are relevant to an anti-access/area-
denial environment. I think it is very prudent to ask the Navy exactly how they
took this guidance and produced a draft RFP that seems to deviate quite substan-
tially from the strategic guidance that has been on the books for years. From what
I can gather via open sources and official statements, there seems to be a view that
organic ISR is the capability gap that can best be addressed by an unmanned system. I totally disagree with that argument. Given what we know about China's modernization path, its stated strategy to deter our forces with long-range and increasingly precise anti-ship ballistic and cruise missiles, and given our clear lack of long-

range and persistent combat strike power from the aircraft carrier, it seems to me the role of civilian policymakers in this process is to ensure that taxpayer dollars are spent wisely. As several of my colleagues and I discussed at the hearing, we believe that applying the disruptive characteristics of unmanned aircraft to the striking power of the carrier air wing is the kind of investment path the Navy badly needs to walk down before it is too late.

Mr. LARSEN. What is the Navy's approach to growth in the draft RFP and what kind of roadmap do you envision going forward?

Admiral GROSKLAGS. Growth beyond threshold capability, which provides an affordable path to ensure system effectiveness against future threats, is a vital part of the UCLASS acquisition strategy. Specific areas of growth, identified and prioritized by Fleet input, include Sensor Payload Adaptability/Modularity, Weapons Capacity, Orbit Capacity, Mission Effectiveness (Survivability), Sustainability, and In Flight Refueling. While some specific growth provisions are required, the RFP also ensures the above growth priorities are adequately prioritized and incentivized, enabling the offerors to propose their best value solutions.

Mr. LARSEN. The first panel expressed concerns with how the Joint Staff, OSD, Navy and Operational Commanders determined the requirements for UCLASS. Please describe the detailed process and reviews that have led to the defined requirements in the UCLASS draft RFP including the organizations involved and the general timeline.

General GUASTELLA. The Joint Staff, Office of the Secretary of Defense, the Navy and Operational Commanders have been involved in requirements definition process for the UCLASS system since 2010. UCLASS has adhered to the Joint Capabilities Integration and Development System (JCIDS) process and has received JROC validation of the UCLASS Initial Capabilities Document (ICD) in June 2011 and the Analysis of Alternatives (AOA) in October 2012. Additionally, several JROC memorandums (JROCM) have been issued further refining requirements and priorities.

The JROC is legislated by Title X U.S.C sec181. Voting members include the Vice Chairman of the Joint Chiefs of Staff (Council chairman) and one general or admiral from the Army, Navy, Air Force, and Marine Corps. Advisory members include, but are not limited to, the Under Secretary of Defense for Acquisition, Technology, and Logistics, the Under Secretary of Defense (Comptroller), the Under Secretary of Defense for Policy, the Director of Cost Assessment and Program Evaluation, and the Director of Operational Test and Evaluation.

The following list captures the history of requirements oversight that has led to the currently defined UCLASS requirements.

June 9, 2011—JROCM 087–11. Approved UCLASS ICD. Directed the AOA address incremental capability growth options and trades to ensure affordability & rapid delivery.

August 11, 2011—Material Development Decision Defense Acquisition Board. Approved AOA commencement stating ''UCLASS is an essential step in the evolutionary integration of unmanned air vehicles into the Carrier Strike Group.''

June 8, 2012—JROCM 086–12. Revalidated UCLASS ICD, prioritizing cost & schedule for an affordable platform in three to six years. Proposed Joint Emerging Operational Need (JEON) for sea-based Intel Surveillance and Reconnaissance not validated.

October 1, 2012—UCLASS AOA Assessment. Director of Cost Assessment and Program Evaluation certified UCLASS AOA for future acquisition decisions.

December 19, 2012—JROCM 196–12. Established refined requirement (e.g., 2 x 24/7 unrefueled orbits at 600 NM, 1 x 24/7 unrefueled orbit at 1200 NM, 2,000 NM unrefueled strike) for an affordable, adaptable platform that supports missions across permissive counter-terrorism and low-end contested environments, enables capabilities for high-end denied operations, and Navy organic missions (Navy incorporated into draft CDD).

April 5, 2013—Draft Capability Development Decision (CDD) reviewed by Navy. Chief of Naval Operations approved Navy UCLASS draft CDD (incorporated JROCM guidance).

April 19, 2013—JROCM 089–13. JROC reviewed overall Joint Unmanned Aerial Systems (UAS) portfolio and refined UCLASS requirements. Approved updated ICD changes since June 2011.

May 21, 2013—JROCM 105–13. Endorsed UCLASS AOA and requestd program update to JROC by 30 Nov 13 to evaluate program against JROCM 196–12 (19 Dec 12).

June 7, 2013—Technical Development Strategy Approved. Under Secretary of Defense for Acquisition, Technology, and Logistics (USD (AT&L)) approved UCLASS acquisition strategy.

June 10, 2013 Congressional Certification. Vice Chairman Joint Chiefs of Staff (VCJCS), Assistant Secretary of the Navy for Research Development and Acquisition (ASN RDA) and USD (AT&L) certified to Congress UCLASS need, program viability, affordability, and compliance with statutes.

October 18, 2013—Executive Requirements and Resources Review Board. Navy leadership reviewed baseline system thresholds, prioritize growth capabilities (i.e., payload adaptability, survivability, weapons capacity, air refueling, sustainability).

October 30, 2013—Force Application Functional Capabilities Board. Navy provided Joint Staff, OSD, Combatant Command, and Service representatives the UCLASS program overview, draft key performance parameters/key system attributes (KPP/KSA), cost, schedule, risks.

November 14, 2013—JROC. Program review of KPP/KSAs, cost, schedule, risks.

February 4, 2014—JROCM 009–14. Endorsed November 2013 review of UCLASS program; established UCLASS Early Operational Capability (EOC) 4–5 years from contract award; emphasized affordability and earliest possible delivery. Directed program update to JROC within 60 days of contract award.

May 22, 2014—Navy update to VCJCS. Program status and draft request for proposal (RFP). Per JCIDS requirements, the UCLASS draft CDD will be revised following technology development and submitted to the JROC for validation prior to Milestone B.